Mouse Phenotypes
A Handbook of Mutation Analysis

Mouse Phenotypes
A Handbook of Mutation Analysis

Virginia E. Papaioannou
Columbia University College of Physicians and Surgeons

Richard R. Behringer
University of Texas M.D. Anderson Cancer Center

COLD SPRING HARBOR LABORATORY PRESS
Cold Spring Harbor, New York

Mouse Phenotypes
A Handbook of Mutation Analysis

© 2005 by Cold Spring Harbor Laboratory Press, Cold Spring Harbor, New York
All rights reserved
Printed in China

Publisher	John Inglis
Acquisitions Editor	David Crotty
Managing Editor	Jan Argentine
Developmental Editor	Judy Cuddihy
Project Coordinator	Inez Sialiano
Production Manager	Denise Weiss
Production Editor	Rena Steuer
Desktop Editor	Susan Schaefer
Book and Cover Designer	Denise Weiss

Front cover: The Garden of Earthly Delights: Allegory of Luxury, central panel of triptych, c. 1500 (oil on panel) by Hieronymus Bosch (c. 1450–1516.) © The Bridgeman Art Library 2001, Prado, Madrid, Spain.

Inside front cover: Flow diagram for analyzing mouse phenotypes using this handbook. Numbers refer to book sections; dashed lines represent dominant mutations.

Title page image: Courtesy of Phil Soriano.

Library of Congress Cataloging-in-Publication Data

Papaioannou, Virginia E.
 Mouse phenotypes : a handbook of mutation analysis / Virginia E. Papaioannou, Richard R. Behringer.
    p. cm.
 Includes bibliographical references (p.).
 ISBN 0-87969-640-0 (printed hardcover : alk. paper)
 1. Mice--Genetics--Handbooks, manuals, etc. 2. Genotype--Handbooks, manuals, etc. 3. Animal mutation--Handbooks, manuals, etc. I. Behringer, Richard. II. Title.
 QH470.M52P37 2004
 599.35'135--dc22
                                                   2004001497

10 9 8 7 6 5 4 3 2 1

Students and researchers using the procedures in this manual do so at their own risk. Cold Spring Harbor Laboratory makes no representations or warranties with respect to the material set forth in this manual and has no liability in connection with the use of these materials. All registered trademarks, trade names, and brand names mentioned in this book are the property of the respective owners. Readers should please consult individual manufacturers and other resources for current and specific product information.

All World Wide Web addresses are accurate to the best of our knowledge at the time of printing.

Procedures for the humane treatment of animals must be observed at all times. Check with the local animal facility for guidelines.

Certain experimental procedures in this manual may be the subject of national or local legislation or agency restrictions. Users of this manual are responsible for obtaining the relevant permissions, certificates, or licenses in these cases. Neither the authors of this manual nor Cold Spring Harbor Laboratory assume any responsibility for failure of a user to do so.

The polymerase chain reaction process is covered by certain patent and proprietary rights. Users of this manual are responsible for obtaining any licenses necessary to practice PCR or to commercialize the results of such use. COLD SPRING HARBOR LABORATORY MAKES NO REPRESENTATION THAT USE OF THE INFORMATION IN THIS MANUAL WILL NOT INFRINGE ANY PATENT OR OTHER PROPRIETARY RIGHT.

Authorization to photocopy items for internal or personal use, or the internal or personal use of specific clients, is granted by Cold Spring Harbor Laboratory Press, provided that the appropriate fee is paid directly to the Copyright Clearance Center (CCC). Write or call CCC at 222 Rosewood Drive, Danvers, MA 01923 (508-750-8400) for information about fees and regulations. Prior to photocopying items for educational classroom use, contact CCC at the above address. Additional information on CCC can be obtained at CCC Online at http://www.copyright.com/

All Cold Spring Harbor Laboratory Press publications may be ordered directly from Cold Spring Harbor Laboratory Press, 500 Sunnyside Blvd., Woodbury, N.Y. 11797-2924. Phone: 1-800-843-4388 in Continental U.S. and Canada. All other locations: (516) 422-4100. FAX: (516) 422-4097. E-mail: cshpress@cshl.edu. For a complete catalog of all Cold Spring Harbor Laboratory Press publications, visit our World Wide Web Site http://www.cshlpress.com/

*To Simon, Chloe, and Gabe
with love*
―――
VEP

To Bob and Fe
―――
RR

A mouse with no phenotype is non-existent
――――――――――
ANNE McLAREN

Contents

Preface, ix

CHAPTER 1   In the Beginning: Obtaining a Mutant, 1

CHAPTER 2   Gene-Targeting Strategies for Maximum Ease and Versatility, 13

CHAPTER 3   Getting to a Phenotype, 29

CHAPTER 4   Maintaining Your Mutant, 49

CHAPTER 5   Phenotypic Analysis: Prenatal Lethality, 71

CHAPTER 6   Phenotypic Analysis: Postnatal Effects, 131

CHAPTER 7   Dominant Effects, 163

CHAPTER 8   Getting Around an Early Lethal Phenotype, 175

APPENDICES   1   Useful Books and Resources, 197
                   2   Examples of Phenotype Analysis, 205
                   3   Glossary, 215

Index, 227

Preface

THE LABORATORY MOUSE, *Mus musculus domesticus*, is the principal mammalian species used in biomedical research as a model organism for human biology. Clearly, one of the central resources for understanding gene control and gene function in vivo is a mutant mouse, and the interpretation of mutant phenotypes is of prime importance in understanding how alterations in the genome result in complex changes in the phenotype. In the postgenome era, the plethora of mouse mutations available is revolutionizing developmental biology.

Suppose you have decided that a custom mutation in your favorite gene would provide you with material to investigate the biological role of that gene and its interactions with other genes in the living organism. You choose to make a mouse model and you have available the powerful methods of molecular biology and targeted mutagenesis that will allow you to precisely alter the DNA in the manner of your choosing. You carefully plan the targeting construct and produce embryonic stem cells with a specific genetic alteration that produces a nonfunctional allele, a dominant negative allele, or any other type of change you desire. But when the new allele is in the context of the whole organism, the mouse, the phenotype is not as straightforward as you had hoped or expected. Where do you start with the analysis? How do you grapple with unfamiliar developmental stages or organs that you have not seen since anatomy class to dissect the phenotype and begin to understand the molecular and developmental consequences of the genotype? How can you make the best use of the valuable resource you have developed and uncover causes of complex phenotypes? If you have acquired a mutation as the result of a spontaneous mutation or a mutagenesis screen, you may be faced with some of the same issues, whether or not you know the molecular basis of the mutation. Here is where we can help to guide you in the basic principles of mutant analysis: How does the mutant animal differ from normal, and when does it begin to differ from normal? This handbook provides the concepts and tools needed to analyze mutant phenotypes in mice, whether they arise spontaneously or are the result of gene manipulations, with an emphasis on mutations that affect embryonic development.

The concept for this book grew out of our experiences teaching the Cold Spring Harbor Laboratory course on Molecular Embryology of the Mouse, or the "mouse course," as it is commonly known. After weeks of learning the language of mouse embryology and sophisticated techniques of gene manipulation, the students at the end of the course still had questions about how to put it all together to analyze a mutant. Their consternation when faced with a complex phenotype resulting from gene manipulation or spontaneous mutation was our biggest challenge, and there was little in the way of reference books we could point to for guidance. This handbook is our attempt to provide a road map of the entire procedure, starting with the planning stages of gene targeting or mutagenesis experiments through the complete analysis of complex phenotypes. Advice on strategy and practical hints will make the analysis of mutant phenotypes a more efficient and productive enterprise.

Even if you are not an expert in mouse development, the simple procedures detailed in this book will allow you to make an intelligent assessment of the phenotype resulting from a mutation and determine the timing and likely causes of specific developmental failures. We provide simple protocols to pin down contributing factors. References to the manual, *Manipulating the Mouse Embryo*, are provided when more detailed protocols are necessary. From there, it is up to you to continue with an in-depth analysis or find an appropriate expert collaborator. Either way, you will certainly gain a greater appreciation for the beauty and complexity of the developing organism, and you may even learn to get over the fear of finding a lethal and learn to love an embryonic phenotype.

This book takes the form of a dichotomous branching pathway of analysis (see the diagram on the inside front cover). By starting at the beginning and answering simple questions about your mutation or situation, you are directed to the most relevant sections of the book, skipping over sections that do not apply. Helpful hints based on our many years of experience will help you to avoid costly mistakes. For specific examples of mutations, methods, and situations, a fully referenced table of examples, drawing heavily on our own experience, is provided in Appendix 2 and is keyed to the book sections in which such topics are discussed.

Many people have helped make this book possible. We would first like to thank co-instructors Peter Koopman, Terry Magnuson, Andras Nagy; assistants Debbie Chapman, Yuji Mishina, Jenny Nichols, and Bill Shawlot; and most of all, the students of the Cold Spring Harbor course on Molecular Embryology of the Mouse, 1995 and 1996, for the inspiration for this book. V.E.P. thanks Robin Lovell-Badge for hosting a sabbatical visit at the National Institute for Medical Research in Mill Hill, London where the inspiration was turned into the beginnings of a book with the support of a Burroughs Wellcome Visiting Professorship. We thank all of our colleagues and students who generously supplied helpful comments or images to illustrate the book, namely, Peter Akinwunmi, Debbie Chapman, Chun-Ming Chen, Zhoufeng Chen, Amalene Cooper-Morgan, Frank Costantini, Sally Cross, Benoit de Crombrugghe, Guy Eakin, Elana Ernstoff, Laurel Fohn, Saadi Ghatan, Jeremy Gibson-Brown, Sarah Goldin, Deborah L. Guris, Kat Hadjantonakis, Zach Harrelson, Akira Imamoto, Soazik Jamin, Loydie Jerome-Majewska, Monica Justice, Robert Kelly, Akio Kobayashi, Kin Ming Kwan, Gigi Lozano, Yuji Mishina, L.A. Naiche, George Ogunrinu, Dmitry Ovchinnikov, Andy Salinger, Gus "T-box" Schama, Reena Shakya, Phil Soriano, and Aya Wada. Thanks are due also to the staff at Cold Spring Harbor Laboratory Press: John Inglis, Jan Argentine, Inez Sialiano, Denise Weiss, Rena Steuer, and Susan Schaefer, with special thanks to David Crotty, our Acquisitions Editor, and Judy Cuddihy, our enthusiastic and ever-patient Editor. And finally, thanks to the members of our labs, past and present, who are the able practitioners of this craft of phenotypic analysis.

VIRGINIA E. PAPAIOANNOU
RICHARD R. BEHRINGER

CHAPTER 1

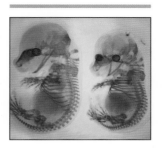

In the Beginning: Obtaining a Mutant

A VARIETY OF WAYS are available to obtain mutations in mice. The most obvious, of course, is to simply wait for spontaneous mutations to arise. But this can be a tedious pastime, and the mutations are unpredictable to say the least. Various other methods have thus been devised either to speed up the process or to make it more predicable. This handbook concentrates on mutations produced by gene targeting in embryonic stem (ES) cells and subsequently introduced into the animal through germ-line transmission from ES cell chimeras. However, the method of phenotypic analysis presented here is by no means limited to this category of mutations; it is equally applicable to mutants produced or collected by other means.

Using gene targeting in ES cells	Go to Section 1.A ▼
Using other means to obtain mutants	Go to Section 1.B ▶

1.A Gene-Targeting Basics

The complex procedures used for targeted gene mutation in the mouse by homologous recombination in ES cells (gene targeting) can be broken down into six basic *gene-targeting steps*.

1. Build a targeting construct specific to your favorite gene.

2. Produce a mutation in ES cells by electroporating the construct that will target a change to one allele of this gene following homologous recombination.

3. Isolate clonal lines of correctly targeted ES cells and grow a sufficient number for further experiments, freezing some for future use.

Chapter opening artwork reprinted, with permission, from Jerome and Papaioannou. *Nat. Genet.* **27:** 286–291 (2001© Nature Publishing Group).

4. Produce ES cell chimeras by injecting the correctly targeted ES cells into mouse blastocysts or by doing morula aggregation.

5. Breed the male chimeras to recover offspring carrying the targeted mutation through germ-line transmission of the targeted allele.

6. Analyze the results of the mutation in heterozygous and/or homozygous mutant mice.

If your starting point is *Step 1*, at the planning stages of building a targeting construct, you can start with *Chapter 2*. If you already have a suitable targeting construct and have electroporated it into germ-line competent ES cells (*Step 2*), it is possible that a mutant phenotype might become evident as early as this point, so our diagnostic procedures start here. You can go directly to *Chapter 3*, which will lead you along a dichotomous branching pathway to get through the rest of the basic *Steps 2–5* to get the mutation into cells, through the germ line of chimeras, and eventually to arrive at a phenotype caused by the mutation. However, before you skip ahead, see *Chapter 2* for useful hints on design of constructs that you might want to review for next time (or in case you need to start again or make another mutation). *Chapter 4* provides a guide to maintaining the mutation once it is safely in the germ line of mice following *Step 5*. Accomplishing *Step 6* can be a long and arduous affair, but it is, after all, the ultimate goal. *Chapters 5–8* provide step-by-step guidance on the methods for efficient analysis of mutant phenotypes regardless of their nature or when they turn up during development.

Before we begin, it is worth taking a moment to reflect on exactly what you expect from your gene-targeting experiment. Ideally, this goal was determined long before the targeting construct was designed, but it is never too late to reconsider your approach. The decision to do gene targeting should not be taken lightly because it is an expensive and time-consuming procedure. Even if everything goes well and works the first time, it will be about six months from the time you have the finished targeting construct ready for electroporation into ES cells until you could possibly see a homozygous mutant animal. In our experience, a year is a much more realistic estimate and two years is not unusual. That being the case, thoroughly thinking the project through and engaging in a few preliminary experiments could save considerable time and trouble.

First, plan the type of mutation that will be most informative and/or versatile. This will depend upon the gene, its products, and how it functions. *Chapter 2* provides some pointers for making the most of the information that you have.

Second, try to predict the phenotype. Although vast experience indicates that the only sure way to determine the effect of a mutation is to make the mutation, this is not a thinking scientist's approach. There are prognosticators that can take some of the guesswork out of the prediction, thus guiding the experimental strategy. Foremost among these is the expression pattern of the gene. Stories circulate about expression patterns that seem to have nothing to do with the mutant phenotype, and mutations that have no phenotype in spite of extensive gene expression. But we maintain that these situations are the exceptions, if they exist at all, and the absence of a phenotype in an expressing tissue is more likely a reflection of our ignorance regarding how to look for it.

A starting point in gene-targeting experiments is thus to define the expression pattern of your chosen gene. It may be your favorite because of its expression in an adult organ of interest or its complicity in a disease process. But remember that the same gene may be utilized during development or in other tissues and could thus have an effect at an unanticipated time and place. Search the literature for reports of the gene's expression and consider doing a survey of expression in a panel of adult organs and embryonic stages, even if only at the level of reverse transcriptase–polymerase chain reaction (RT-PCR) or northern blot analysis (commercial blots are available). Although a detailed expression analysis may not be what you choose to do at this stage, expression in embryos will at least alert you to the possibility of an embryonic phenotype. You can also check the expressed sequence tag (EST) databases to identify tissues in which transcripts for your favorite gene have been isolated as a guide to expression.

Armchair biology can get you into the right frame of mind for dealing with the unexpected by anticipating some of the possibilities of a gene disruption. With the current amount of genome information available, it is possible to learn a lot about your gene that might bear on its function. The following is a checklist of what you should know about the gene to help you prepare before you begin.

- *What is the chromosomal location of your gene?* This is an important piece of information for many reasons. Is the gene X linked? If so, males will be hemizygous for the mutation and effects might show up in XY ES cells, chimeras, or males, where there is only a single X chromosome. Heterozygous females, which will have one normal allele active in half of their cells due to X inactivation, might escape this problem, unless the gene is expressed in the extraembryonic tissues that are subject to imprinting effects (see Box 1.1 and Figure 1.1). Alternatively, they might show mosaic effects reflecting the mosaicism of cells with the mutant allele on the active X chromosome. Is the gene Y-chromosome linked? Similar to X linkage, there will be only one copy of the gene in XY cells.

- *Does the gene fall within a chromosomal region that is subject to imprinting effects?* If so, heterozygous offspring may be differently affected depending on the parent of origin of the mutant allele (Figure 1.2).

- *Are there known mutations that map to the chromosomal region in which your gene resides?* Check these carefully because if they have not yet been cloned, they could turn out to be alleles of your gene. To see if this is the case, you might want to obtain DNA or the mutants themselves.

- *Is your favorite gene a member of a gene family?* If family members overlap in their expression pattern, some overlap of function could lead to redundancy. In this case, the areas of unique expression of your gene might indicate areas in which a mutant effect is most likely.

- *Is anything known about orthologous or homologous genes in other species?* Look for any existing mutations in your gene in other species or any chromosomal deletions in humans that include this gene. Determine if any human disease loci are mapped to the chromosomal region. Any of this information could provide a clue as to the expected phenotype.

- *Can you make predictions based on the gene expression pattern and the nature of the protein?* Expression patterns tell you when and where the action of a mutation could have its origins. Proteins can act within the cells in which they are produced (cell autonomous) and/or they can act on other cells (cell nonautonomous).

- *Is anything known about interactions of the gene product with other gene products?* This information may be useful for placing the gene product into known biochemical pathways.

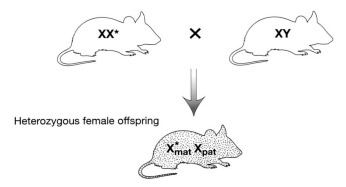

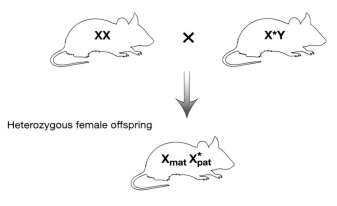

Figure 1.1. Imprinting of an X-chromosome-linked mutant gene. If an X-linked mutant allele (*) is inherited from the mother (*top cross*), heterozygous female offspring are effectively null for the gene product in the trophectoderm and primitive endoderm lineages because the paternally derived X chromosome (X_{pat}) is preferentially inactivated and the maternally derived chromosome (X^*_{mat}) is active. If the mutant allele is inherited from the father (*bottom cross*), heterozygous females are wild type for the gene in the extraembryonic tissues because the maternally derived allele is the only one active in the trophectoderm and primitive endoderm lineages.

All of these considerations, which can lead to fruitful lines of investigation, are better thought through in advance before embarking on a year's worth of gene targeting. In addition, consider that the mutation may result in a mouse that looks perfectly normal. Prior knowledge about the gene and gene products could help reveal a subtle or cryptic phenotype.

With these cautionary words in mind, you are now ready to move on to *Chapter 2* if you are in the process of planning a targeting construct or *Chapter 3* if you have already produced a targeting construct.

Planning a targeting construct	Go to Chapter 2 ▸
Targeting construct already in place	Go to Chapter 3 ▸

Box 1.1 **Imprinting and X Inactivation: Effect on Embryos Due To Preferential X Inactivation of Paternal X Chromosome in Extraembryonic Tissues**

As a rule, X inactivation results from random inactivation of one or the other X chromosome so that females end up with a random mix of cells with either the paternally derived X or the maternally derived X inactive. However, there is an exception to the random inactivation rule in the embryo: There is preferential inactivation of the paternally derived X chromosome in the trophectoderm and primitive endoderm, and consequently in their cellular derivatives. This peculiarity of the embryo, which is a form of imprinting, usually has little effect on experimental work. However, if you have produced a mutation in an X-linked gene that is expressed in either of these tissue lineages, e.g., placental structures or yolk sac endoderm, there will be two classes of heterozygous females with respect to gene expression. One class will be functionally null in the trophectoderm and primitive endoderm lineages because it inherits the mutant allele from the mother and the paternal chromosome is inactivated. The other class will be functionally wild type because it inherits the mutant allele from the father and the maternal chromosome is active (Figure 1.2).

1.B Other Means to Obtain Mutants

The spontaneous mutation rate in mice is one gamete in 100,000 for a specific gene and, of course, the identification of spontaneous mutations is biased toward mutant phenotypes that can be detected visually (i.e., coat color, gross morphology, and neurological behavior). Attempting a genetic screen for spontaneous mutations in mice is thus logistically impractical. However, in addition to gene targeting in ES cells, there are many other ways to induce mutations in mice. These methods can be grouped into broad categories, including chemically induced mutations, radiation-induced deletions, insertional mutations, and transgene expression. With these methods, mutations in specific genes of interest can be either selected or directed.

1.B.1 Chemical mutagens

Chemical mutagens can greatly increase the mutation frequency, making genetic screens in mice highly feasible. Recently, a number of groups around the world have performed small- and large-scale chemical mutagenesis screens in mice to isolate dominant and recessive mutations. The most powerful chemical mutagen in mice is ethyl nitrosourea (ENU) (see Box 1.2). ENU predominantly causes point mutations, creating nonsense, missense, splice, and regulatory mutations that can affect gene, RNA, and protein function; null, neomorph, antimorph, hypomorph, and hypermorph mutations can thus be isolated. ENU induces an average per-

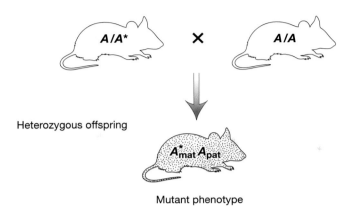

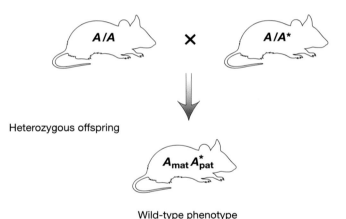

Figure 1.2. Inheritance of an imprinted, autosomal mutant gene. In this example, the locus is only expressed from the maternal allele. When a mutation (A^*) is inherited from the mother (*top cross*), heterozygous offspring will have one mutant maternal allele (A^*_{mat}) and one inactive paternal allele (A_{pat}) and will thus show a mutant phenotype. When the mutation is inherited from the father (*bottom cross*), the wild-type maternally inherited allele will be the active copy and the heterozygous offspring will be phenotypically wild type. It should be noted that the situation would be reversed if the imprinted locus is expressed from the paternal allele.

locus mutation frequency of 1.5×10^3. A mutation in a single gene of choice can be recovered on average in one in 600–700 gametes screened. To date, every ENU-induced mutation that has been sequenced in the mouse is a point mutation and, because ENU causes point mutations, it defines single gene function. The advantages of ENU mutagenesis are the rich variety of alleles generated, the possibility of identifying mutations by phenotype during the screen, and the fact that many mutations can be generated and isolated using simple methods (Box 1.3). The primary disadvantages are the mouse space and labor required to perform a screen and the effort required to map and identify the molecular lesions.

> **Box 1.2 ENU Mutagenesis**
>
> Typically, adult males of a particular inbred strain are injected intraperitoneally with a precise dose of ENU to mutagenize spermatogonial stem cells. ENU treatment causes a depletion of mature germ cells, resulting in transient sterility. The spermatogonial stem cells are spared but their DNA acquires numerous point mutations throughout the genome, depending on the dose of ENU administered. The spermatogonial stem cells then undergo spermatogenesis to produce sperm. The ENU-injected males usually regain fertility (11–17 weeks post-ENU treatment) and are then bred using different breeding schemes designed to generate pedigrees for phenotypic analysis to isolate mutations.

A phenotypic screen will identify mutations affecting specific structures or processes of interest. However, if mutations in a specific gene are desired, an alternative is to screen a frozen sperm archive. The sperm of G_1 mice derived from males that have been mutagenized with ENU are cryopreserved and their tissues saved. These tissues can be used as a source of DNA to screen for mutations in specific genes using PCR, denaturing high-performance liquid chromatography, or other methods to detect mismatches. Once a mutation is identified, the sperm can be thawed and used to recover mice with the mutation.

ES cells can also be mutagenized with ENU to create point mutations and then screened for mutations in specific genes. cDNA or genomic DNA isolated from pools of ENU-mutagenized ES cell clones may be analyzed by PCR or RT-PCR for mutations in specific genes. Once a clone carrying a mutation in the specific gene is identified, it can be used to generate mice carrying the mutation through standard techniques for making ES cell chimeras and recovering offspring with the mutation.

> **Box 1.3 Genotyping ENU Alleles**
>
> Mice with fully penetrant dominant mutations that are viable and fertile are easy to genotype because the dominant phenotype can be used to identify the mutants. However, mice heterozygous for recessive ENU-induced mutations can initially be difficult to genotype because the gene and chromosomal location are usually unknown. In these cases, one must perform test crosses to identify carriers (Figure 1.3). Genetic mapping eventually leads to the identification of the chromosomal location of the mutation. Subsequently, linked polymorphic markers can be used to track the mutation. Ultimately, the ENU mutation is identified and, if you are fortunate, it may cause a restriction enzyme polymorphism that can be exploited in a PCR genotyping strategy. Alternatively, if there is no restriction enzyme polymorphism, single-stranded conformation polymorphism or other strategies can be used for genotyping.

1.B.2 X-ray mutagenesis

Exposing mice to X rays causes chromosomal deletions in germ cells that can be transmitted to progeny. The mutagenized mice are bred to mice with appropriate genetic markers to identify the

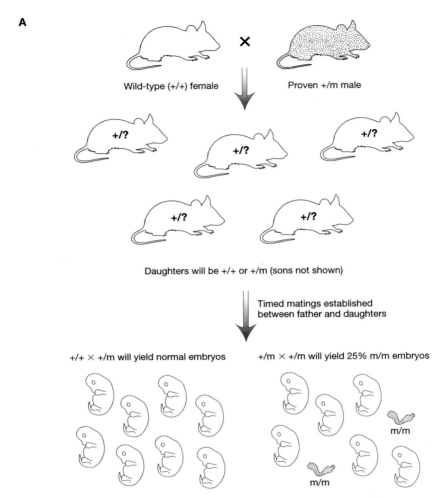

Figure 1.3. Test crosses for recessive ENU-induced mutations. Spontaneous and ENU-induced mutations may not be molecularly defined. In this case, genetic crosses must be performed to identify mice that carry the mutation. Two situations are shown. (*A*) A male known to be heterozygous for a recessive embryonic lethal mutation is bred with wild-type females to generate progeny. The daughters from this cross will be +/+ or +/m. To obtain homozygous mutant embryos for analysis, timed matings are established between the male carrier and his daughters for dissection at time points chosen to yield the mutant phenotype. Approximately 50% of the time, the cross will be between a +/+ daughter and the +/m father that will yield +/+ and +/m embryos having a wild-type phenotype. The other 50% of the time, the cross will be between a +/m daughter and the +/m father that will yield +/+, +/m, and m/m embryos. Therefore, ~25% of the embryos from this cross should show the mutant phenotype. If new heterozygous males are needed, then the sons from a known carrier are test crossed with their siblings to distinguish +/+ and +/m males. (*Continued on facing page.*)

deletion alleles. Classically, this approach has been used to generate deletion complexes, which usually include many loci. One well-characterized deletion complex is the albino deletion complex, a series of overlapping deletions that include the *albino* (*Tyr*) locus. These deletions are identified by test crosses with mice that are homozygous for the recessive *albino* allele (*Tyr^c*). The presence of white mice in the progeny indicates that the *albino* locus was deleted in the germ cells

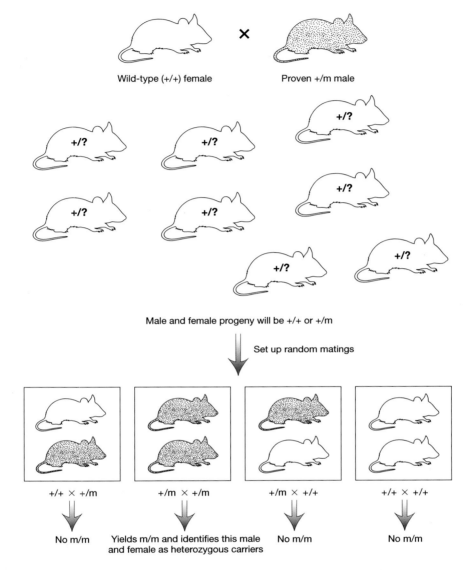

Figure 1.3. (*Continued from facing page.*) (*B*) A male known to be heterozygous for a recessive viable but visible mutation is bred with wild-type females to generate progeny. The offspring from this cross will be +/+ or +/m. To obtain homozygous mutants for analysis, matings are established at random between the siblings. There is thus a one in four chance that you have set up a mating between two +/m mice. Once progeny are born, they can be examined for the mutant phenotype. Those cages yielding homozygous mutants provide the genetic proof that both of the parents are heterozygous carriers.

of the X-ray-treated parent. By combining various overlapping deletions, one can define loci that affect diverse developmental processes. These genetic resources tend to be used primarily for gene discovery rather than as primary tools for understanding the function of one particular gene.

As with ENU, ES cells can also be irradiated with X rays to induce deletions and then used to generate mutant mice. A system has been developed to target specific chromosomal regions for X-ray-induced deletions: A herpes simplex virus thymidine kinase (HSV *tk*) gene is introduced into a specific locus by gene targeting in ES cells. The ES cells are then irradiated and the loss of HSV *tk* is selected for by growth in ganciclovir- or FIAU (1,2-deoxy-2-fluoro-1-β-D-arabinofuranosyl-5-iodouracil)-containing selective medium. Clones that survive will have deletions that include the introduced HSV *tk* gene and an unknown amount of the targeted endogenous gene. F_1 hybrid ES cell lines are used so that DNA polymorphisms can be exploited to map the deletion end points. Using this method, many deletions of a specific chromosomal region can be isolated quickly. One drawback is that you cannot control the size and boundaries of the deletions.

1.B.3 Insertional mutagenesis

One of the by-products of generating transgenic mice by pronuclear injection of DNA into zygotes is insertional mutation. It is estimated that approximately 5–10% of random integrations of foreign DNA into the genome fortuitously disrupt an endogenous gene. The transgene can serve as a molecular tag to clone the disrupted gene(s). One of the problems with transgene insertional mutations is that they can be associated with large deletions, duplications, inversions, and translocations. It is thus not always clear if the mutant phenotype is caused by one or multiple gene disruptions. If chromosomal rearrangements are present at the site of transgene integration, molecular cloning of the gene responsible for the mutant phenotype can be difficult. Another potential complication is that the expression of the transgene itself may confuse the analysis of the insertional mutant phenotype.

Viruses can be used to infect preimplantation-stage mouse embryos to cause insertional mutations. As with transgene insertional mutations, the retrovirus can be used as a probe to follow proviral inserts in pedigrees and ultimately clone mutated genes. Retroviruses can also be modified for specific mutagenesis screens (for example, gene traps).

Gene traps are DNA or viral vectors typically introduced into ES cells by electroporation or retroviral infection to generate insertional mutations that disrupt the function of endogenous genes. Gene trap vectors are engineered to include reporters such as *lacZ*, which, in principle, allow you to follow the expression pattern of the trapped locus. ES cell lines with gene trap insertions can be prescreened for reporter expression patterns or to identify the trapped genes. Ultimately, mice generated from ES cell lines with gene traps can then be screened for mutant phenotypes. Because gene traps are essentially random insertional mutations, the resulting mutant alleles can have diverse activity. It is easy to genotype heterozygous mice for gene trap insertions because the vector serves as a dominant molecular marker, but identification of homozygous mutant mice is difficult until the locus is cloned. Libraries of ES cell lines with gene trap insertions and associated sequence tags are available (see Appendix 1). These resources can facilitate the generation of a mutant for a specific gene by allowing you to simply acquire the mutant ES cell clone and generate mice from ES cell chimeras.

Transposons are another class of insertional mutagens that can be used in mice to generate mutant phenotypes. Sleeping Beauty (SB) is a Tc1/mariner-type transposon resurrected from salmonid fish and developed as a transposon mutagenesis system in mice. The SB transposase specifically recognizes SB transposon inverted repeats. The SB transposon is initially introduced into the mouse genome as a transgene either by pronuclear injection or through transfection of ES cells. Transgenic male mice carrying the SB transposon transgene are then bred with transgenic mice that express the SB transposase in the germ line. The resulting double transgenic mice are then bred with wild-type mice and mobilization of the transposable element is monitored in the next generation. The SB transposase causes the transposition of the SB transposon through a precise cut-and-paste mechanism to TA dinucleotides. This transposition can occur in *cis* locally on the same chromosome or in *trans* to other chromosomes. SB is currently being modified to include visual markers of transposition and conversion into gene traps. One advantage of this system is that the mice create more mutations simply by breeding. One outstanding question is the mutation frequency relative to other approaches.

1.B.4 Transgenics

Finally, transgenic mice generated by various methods are frequently created to express a foreign gene product to elicit a mutant phenotype. These are typically gain-of-function experiments for the intentional misexpression or overexpression of a gene in a specific tissue and are different from transgene-induced insertional mutations that randomly disrupt endogenous genes. By design, a tissue-specific promoter may have been used to direct the expression of the gene, so one already knows which tissues to examine for a phenotype in the transgenic mouse. However, sometimes the resulting phenotypes are not as simple as expected and a more comprehensive analysis will be necessary.

With a mutation of your favorite gene in hand, however you obtained it, you can now move on to the phenotypic analysis with a slight detour through *Chapter 4* to get pointers on breeding and maintaining your mutant.

Are heterozygous mutants normal and fertile?

YES	Go to Chapter 5 ▶
NO	Go to Chapter 7 ▶

CHAPTER 2

Gene-Targeting Strategies for Maximum Ease and Versatility

THE MOLECULAR TECHNIQUES USED TO BUILD A GENE-TARGETING CONSTRUCT are standard (see Appendix 1) and the goal of producing a mutation in a specific gene is conceptually straightforward. However, many aspects of the strategy and design of the targeting construct will have a bearing on the success of the procedure and on the versatility of the resulting mutation. With little additional work, you may be able to build features into the initial targeting vector that will save you from having to construct a second targeting vector as the project unfolds. The strategies discussed in this chapter will help you anticipate future needs and provide you with information to evaluate the trade-offs inherent in designing different kinds of alleles. Before you begin, make sure that you already have thorough and detailed information about the gene you intend to target (see Box 2.1) to facilitate the best design for the mutation.

2.A **Gene-Targeting Steps 1 and 2.** *Build a targeting construct specific to your favorite gene and produce a mutation in embryonic stem (ES) cells.*

Presented are general design considerations for all types of alleles using

- homologous DNA,
- positive selection for vector incorporation,
- negative selection against random integration events, and
- screening for homologous recombination events.

Chapter opener artwork courtesy of Robert Kelly.

Box 2.1 Gene Checklist Before Constructing Targeted Mutant Alleles

Answers to the following fundamental questions about the locus in question will facilitate the best design for your desired mutation.

- How many coding and noncoding exons are present?
- Where is the start of transcription?
- Which exon contains the translation start codon?
- Which exon contains the translation stop codon?
- How many kilobases does the genomic region encoding the open reading frame span?
- Is there evidence for alternative splicing? If so, which exons are alternatively spliced?
- Are protein isoforms generated by the locus?
- Can you easily delete the entire locus, i.e., can the entire locus be deleted by removing <20 kb?
- If you delete a specific exon(s), can the remaining exons splice together in-frame?
- If the gene (genomic region) is relatively large (>20 kb), is there a group of contiguous exons that reside within a small (<20 kb) genomic region that could be deleted?
- Do you have an accurate restriction map of the locus for the specific strain of the ES cells?
- Do you have a sequence of the locus from the strain of mice from which the ES cells were derived?
- Have you performed a sequence comparison of nonexonic sequences among human, mouse, and rat to identify the location of conserved sequences that might regulate expression?

2.A.1 Homologous DNA

Certain details should be considered when generating any type of gene-targeting vector. One of the most important is to use genomic DNA that is isogenic with the ES cell line that will be used for the gene-targeting experiments; in other words, both the genomic DNA clones and the ES cells should be derived from the same inbred strain of mouse. A complete sequence match between the targeting vector homology and the locus to be mutated facilitates obtaining the maximum frequency of homologous recombination. Conversely, potential sequence mismatches resulting from the use of DNA from a different mouse strain can reduce the frequency of homologous recombination or even prevent it altogether.

The total amount of sequence homology to be used in the gene-targeting vector should be between 5 and 8 kb, which represents a balance between increased chances of homologous recombination and ease of vector construction: Smaller amounts of homology can reduce the targeting frequency and larger amounts can complicate vector construction as well as the identification of homologous recombinants. In general, the homologous sequence should be evenly split between the upstream and downstream arms of homology in the vector (Figure 2.1). However, if necessary, because of gene structure or cloning constraints, one arm of homology can be as short as 1 kb in length with the other arm making up the balance of the total recommended homology of 5–8 kb. The arms of homology in the targeting vector should flank any genomic sequences you wish to delete.

Gene-Targeting Strategies for Maximum Ease and Versatility | 15

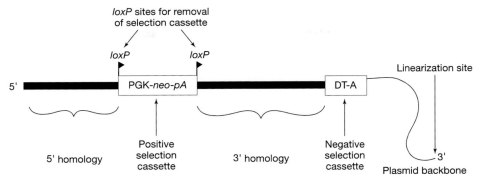

Figure 2.1. Elements of an idealized targeting construct. The targeting constructs used for different targeting strategies vary in detail, but the usual components necessary for successful gene targeting include (1) homologous, isogenic DNA totaling 5–8 kb and evenly split between upstream and downstream arms of the targeting construct; (2) a positive selectable marker, such as *neo*, driven by a promoter active in ES cells, such as PGK, and surrounded by *loxP* sites to facilitate its eventual removal from the targeted allele; (3) a negative selectable marker, such as DT-A, to screen against random integration events, located outside the region of homology; and (4) a linearization site in the plasmid backbone located some distance from the negative selectable marker to protect it from degradation of the ends after linearization. The other essential elements (not shown) are 5′ and 3′ probes to the endogenous locus that are external to the regions of homology in the construct. These are used in Southern analysis of restriction enzyme digests to diagnose 5′ and 3′ crossover events to recognize a correctly targeted allele.

2.A.2 Positive selection for vector incorporation

Positive selection of ES cell clones that have integrated the targeting vector is usually achieved using a neomycin phosphotransferase (*neo*) gene expression cassette that confers resistance to the drug G418, although other selectable markers can be used. However, the presence of a selectable marker gene with its associated promoter (e.g., the PGK promoter) at targeted loci has been shown in some cases to influence the expression of neighboring genes. Therefore, it is a good idea to flank the positive selectable marker by *loxP* sequences (this is known as a "floxed" allele; see Section 2.A.5), providing the option of removing the selectable marker later with Cre recombinase. Once homologous recombination has been successful, the floxed selectable marker can be deleted either in the targeted ES cells or later in mice. Removal is accomplished in ES cells by transient in vitro Cre expression before advancing to the next step of making chimeras. If mice carrying the targeted allele have been produced, the floxed selectable marker can be deleted by pronuclear injection of a Cre expression plasmid into zygotes carrying the floxed gene or by breeding mice carrying the floxed gene with Cre transgenic mice (see Chapter 4, Section 4.D.1). In either case, using a floxed allele strategy provides a simple way to generate two different alleles for the targeted gene, one with the selectable marker and one without. Having two different alleles could be useful for chimera studies later in the analysis of a phenotype (Chapters 5 and 8).

An added benefit of the floxed selectable marker strategy is that it provides the opportunity to make homozygous mutant ES cell lines without ever making a mouse. By deleting the selectable marker from the targeted locus in ES cells, you can reuse the original targeting vector and the same selective drug a second time to target the remaining wild-type allele, a strategy called "marker recycling."

2.A.3 Negative selection against random integration events

Negative selection against ES cell clones that have incorporated the targeting construct by random insertion can be accomplished using a herpes simplex virus (HSV) thymidine kinase (*tk*) gene outside the region of homology in the construct and culture of the electroporated ES cells with ganciclovir or FIAU (1,2-deoxy-2-fluoro-1-β-D-arabinofuranosyl-5-iodouracil). Cells that contain the *tk* gene resulting from random integration of the entire vector will be sensitive to drug selection (see Box 2.2). Negative selection can also be achieved using a diphtheria toxin A-chain (DT-A) gene that does not require the addition of a drug for selection, but rather kills all cells in which it is incorporated and expressed.

2.A.4 Screening for homologous recombination events

Even after positive-negative selection, many of the ES cell colonies that survive the drug selection will not be homologous recombinants but rather, will contain random integration events where the negative selectable marker was fortuitously lost or damaged. Therefore, all resistant colonies must still be screened using Southern or polymerase chain reaction (PCR) analysis to identify ES cell clones with correct gene targeting. Southern analysis requires the use of a diagnostic restriction enzyme digest with a probe, called the external probe, that is not contained within the regions of homology included in the targeting vector. A unique restriction enzyme site is usually introduced along with the positive selection marker during construction of the targeting vector for ease of screening. The diagnostic restriction enzyme site should be located outside of any floxed selectable marker so that after Cre expression and deletion of the marker gene, the diagnostic restriction enzyme site remains. Using this strategy, a diagnostic digest and hybridization with an external probe will yield a smaller DNA fragment for the mutant allele compared to the wild-type allele. This is desirable for technical reasons because if the mutant band were larger, it would be hard to distinguish from a partial restriction enzyme digest on a Southern blot.

Box 2.2 HSV *tk* and Male Sterility

HSV *tk* expression is toxic to the germ cells of male transgenic mice and causes dominant sterility. This was initially observed in transgenic mice that fortuitously carried an HSV *tk* sequence as part of a larger transgene construct. Subsequent studies revealed that the coding region of HSV *tk* contains a promoter that is transcriptionally active in the differentiating germ cells of male transgenic mice. Fortunately, HSV *tk* is not active during female gametogenesis, and HSV *tk*-containing transgenes can be transmitted through the female germ line. HSV *tk* is widely used in gene-targeting strategies and if it is retained at the targeted locus, it can cause unanticipated difficulties in achieving germ-line transmission of the mutant allele from male chimeras. To get around this problem, the male germ cell promoter of HSV *tk* has been mutated to create a variant *tk* protein that retains enzyme activity for negative selection but is not expressed in male germ cells and is therefore compatible with male fertility.

To characterize the targeted allele fully—and this is essential—both 5′ and 3′ external probes must be used to decipher the structure of the gene-targeting events that have occurred through both the 5′ and 3′ arms of homology. If for some reason both 5′ and 3′ external probes cannot be identified, the targeting event can be identified initially using one external probe, and the allele can be further investigated by additional analysis using a probe within the region of targeting vector homology (e.g., an internal probe such as *neo*).

2.A.5 Site-specific recombinases

Site-specific recombinases have become standard tools for making targeted mutations in mice. The primary system currently used is the Cre/*loxP* system but the Flp/*FRT* system is also used. The ΦC31 integrase/*attPP′* and *attBB′* system is a new application to gene targeting in ES cells, but it may prove useful for sophisticated gene alteration strategies. Cre is a DNA recombinase that specifically recognizes 34-bp sequences called *loxP* sites. *loxP* sites can be placed on either side of a segment of DNA. The length of the intervening DNA can be relatively small

---

Box 2.3   Removing a Selection Cassette from ES Cells In Vitro

To remove the selection cassette from your targeted ES cell clones in vitro, transiently express the appropriate recombinase (usually Cre or Flp) and then isolate clones and genotype for the removal of the cassette. Electroporate circular plasmids into the ES cell clones, plate at clonal density, and then culture without selection. Pick and then genotype the ES cell clones. Usually, about 10–20% of the clones will have had the cassette removed by the transient recombinase expression. In the past, some investigators have used a floxed *neo*-HSV *tk* dual cassette so that the loss of *tk* can be selected for by culture in gancyclovir or FIAU. This results in nearly all of the surviving ES cell clones having the floxed cassette removed. If you do this, however, remember that HSV *tk* expression causes male sterility (see Box 2.2) and male chimeras made with ES cells that retain a *neo*-HSV *tk* will likely be sterile.

Triple *loxP* strategies have traditionally been used to generate conditional null alleles. In this strategy, two *loxP* sites flank a selection cassette (e.g., *neo*) and the third *loxP* site is added to flank important exons. To remove the selection cassette in vitro, it is necessary to express sufficient Cre to remove the floxed cassette, but not so much as to delete the floxed exons. This can be tricky and may require titrating the amount of Cre expression plasmid used for electroporation or using a Cre plasmid that is less active. In theory, you should be able to isolate four types of ES cell clones after transient expression of a Cre plasmid in your triple *loxP* targeted ES cell clones: (1) triple *loxP* intact; (2) all floxed sequences deleted (null); (3) deletion of the floxed exons (null), but retention of the floxed selection cassette; and (4) deletion of the floxed selection cassette, but retention of the floxed exons—your conditional null allele. Genotype these clones very carefully because they may be mosaic for these four possible outcomes, depending on when Cre acted as the cell generating the clone proliferated after electroporation. Nowadays, many labs use a dual Cre/Flp strategy to flox important exons and flrt the selection cassette. Thus, expression of Flp will remove the selection cassette, leaving the important exons floxed.

One caveat of the in vitro approach for removing a selection cassette is that it requires a second electroporation and cell cloning step that could compromise obtaining germ-line transmission of the targeted allele. In addition, it is necessary to choose the targeted clone(s) to electroporate with the recombinase vector. Often, only one clone is chosen for this purpose, which can be a big mistake because that clone has probably not been screened for germ-line transmission ability after the initial targeting. An alternative is to remove the selection cassette in vivo, after mice have been produced (see Section 4.D.1).

(<1 kb) or very large (cM). The central 8-bp core of the *loxP* site is flanked by inverted repeats and specifies an orientation to the 34-bp sequence. When *loxP* sites are placed on either side of a piece of DNA, the intervening DNA is called "floxed" (flanked by *loxP* sites). Because the *loxP* sites have an orientation, they can flank a piece of intervening DNA in either direct orientation or inverted orientation. Cre will recognize *loxP* sites flanking a piece of DNA and catalyze the deletion of the intervening DNA if the flanking *loxP* sites are in direct orientation or it will invert the intervening DNA if the *loxP* sites are in inverted orientation. A similar situation occurs with Flp recombinase and *FRT* sites. DNA sequences flanked by *FRT* sites have been called "flrted."

Generally, selectable marker cassettes are now floxed or flrted so that they can be removed from the targeted locus by Cre or Flp, respectively. In this way, extraneous DNA sequences that might complicate the analysis of the mutant are removed from the targeted locus. Elimination of the selectable markers can be performed in ES cells (Box 2.3) or in mice through crosses with mice that express the recombinases (see Section 4.D.1).

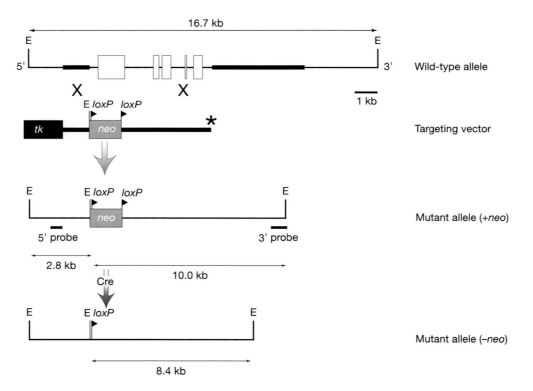

Figure 2.2A. Deletion of all protein-coding exons for a small gene. All of the protein-coding exons of the wild-type allele (*open boxes*) are replaced by a floxed neomycin (*neo*) cassette to ensure the generation of a null allele. In this case, the total amount of homology in the targeting vector (5.5 kb) is divided as a 1.2-kb 5′ arm and a 4.3-kb 3′ arm (*thick lines*). The *neo* cassette includes an *Eco*RI (E) site to identify homologous recombinants by Southern analysis. An HSV thymidine kinase (*tk*) cassette is placed adjacent to the 5′ arm of homology for negative selection. The site of vector linearization is indicated by the asterisk (*). In this case, 5′ and 3′ external probes are used to identify the targeted mutant allele. The sizes of the wild-type and mutant bands for Southern analysis are indicated. The floxed *neo* cassette can be removed by Cre. (+*neo*) Allele retaining the *neo* cassette; (−*neo*) allele in which the *neo* cassette has been deleted.

2.B Different Types of Alleles

2.B.1 Making a null allele

The first mutation most people aim to make when they initiate the study of a particular gene is a complete loss of function, i.e., a null allele (see Figure 2.2). Null alleles are very useful because the phenotype of the homozygous null mouse provides information about the earliest essential role of the product(s) expressed from the locus during development. In addition, null alleles can be used in combination with other alleles for more complex studies. Fortunately, a null allele is one of the simplest mutations to generate, both conceptually and technically, by gene targeting in ES cells.

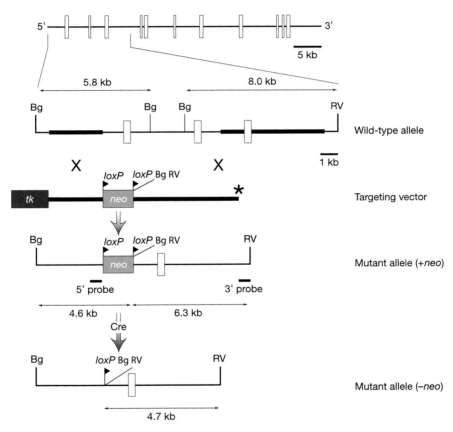

Figure 2.2B. Deletion of the initial protein-coding exons for a large gene. The coding exons (*open boxes*) of this gene span 38 kb (*top*). Therefore, the first two coding exons are deleted. This region (6.3 kb) is replaced by a floxed *neo* cassette. A total of 9.0 kb of homology is divided into a 3.0-kb 5′ arm and a 6.0-kb 3′ arm (*thick lines*). The *neo* cassette includes *Bgl*I (Bg) and *Eco*RV (RV) sites to identify homologous recombinants by Southern analysis. An HSV *tk* cassette is placed on the 5′ arm of homology. The site of vector linearization is indicated by the asterisk (*). A 3′ external probe is used for the initial identification of homologous recombinants. Once these clones are identified, they are expanded and can be analyzed with an internal probe to characterize the structure of the 5′ recombination event. The sizes of the wild-type and mutant bands for Southern analysis are indicated. The floxed *neo* cassette can be removed by Cre. (+*neo*) Allele retaining the *neo* cassette; (*neo*) allele in which the *neo* cassette has been deleted.

For small genes (≤20 kb), the most straightforward strategy is to delete all protein-coding exons (Figure 2.2A); this guarantees the generation of a null allele because protein products cannot be synthesized from the targeted locus. Thus, once the deletion of the DNA sequences has been verified, it is not necessary to analyze mRNA or protein for the targeted gene in homozygotes, because there will be none expressed. If the coding sequences are not removed in their entirety, mRNA and protein could still potentially be expressed from the targeted locus, and characterization of variant transcripts generated from the targeted allele will be necessary. This characterization can be time-consuming and difficult to interpret. Partial protein products that could potentially be synthesized from a targeted locus may be problematic if they retain some function. Antibodies for the gene product may not be available to determine if a partial protein product is being expressed, further complicating the interpretation.

An argument against designing a deletion to remove all coding exons would be the presence of regulatory elements for neighboring genes located within the introns. This possibility should be kept in mind, especially if other genes are located nearby. Sequence comparisons of genomic DNA between mouse and human will reveal if there are conserved sequences within the introns that might indicate the presence of such regulatory elements, but they will not necessarily reveal the identity of the gene being regulated.

Generating null alleles for a larger gene requires a little more thought because standard gene-targeting strategies are not as efficient for generating large (>20 kb) deletions; thus, more complex strategies are necessary (Figure 2.2B). With the deletion of all protein-coding exons of a large gene, it is even more likely that regulatory elements of adjacent genes, or even all or part of another gene, could be located within the introns of the gene being deleted. Sequence comparisons of genomic DNA between mouse and human should reveal conserved sequences within introns, and searches of expressed sequence tag (EST) databases can indicate if another gene is within or overlaps with the gene to be deleted.

With larger genes, the strategy most likely to result in a null allele is the creation of a deletion of up to 20 kb that removes the initial protein-coding exons, including the exon containing the translation initiation site. If known, the transcription initiation site(s) can also be deleted. If a deletion is designed to remove a subset of internal protein-coding exons, it is important to check that the exons that remain after the deletion cannot splice in-frame to generate a partial protein product. If they can, reconsider the position of the deletion or introduce a frameshift mutation.

Deletions for mutating either large or small genes are created most simply by using a replacement gene-targeting strategy: A selectable marker replaces coding exons and a positive-negative selection scheme is used to enrich for ES cell clones that have undergone homologous recombination. The primary features of a replacement vector are a plasmid backbone containing a positive selection marker positioned between two regions of homology and a negative selectable marker next to one of the arms of homology (see Figure 2.1). It is not essential to use a negative selection marker for gene targeting, although its inclusion will increase the ratio of targeted to randomly integrated clones, thus cutting the workload during screening of drug-resistant ES cell clones. Linearized targeting vectors are usually introduced into ES cells by electroporation, followed by culture in selective medium, and subsequent identification of homologous recombinant ES cell clones (MM3).*

---

*MM3 refers to *Manipulating the mouse embryo: A laboratory manual*, 3rd edition (Nagy et al. [2003], Cold Spring Harbor Laboratory Press, Cold Spring Harbor, New York).

2.B.2 Reporter gene knock ins

On many occasions during phenotypic analysis of a mutant allele, it is useful to have a simple cellular reporter to follow the expression of the gene or to track the fate of mutant cells. Introducing a reporter into a specific locus by gene targeting (a so-called knock-in strategy; see Figure 2.3) can be used to achieve the goal of faithful recapitulation of the expression of an endogenous gene by a readily detectable reporter. This method exploits the intact regulatory elements of the endogenous chromosomal locus, at the same time rendering the gene nonfunctional by deletion or alteration of crucial protein-coding exons. One of the important steps in designing a knock-in allele is to determine the position within the gene to introduce the reporter. The reporter gene can be engineered (1) as an ATG fusion with the endogenous transcript, (2) as a fusion protein by generating a bicistronic transcript using an internal ribosome entry site (IRES) sequence, (3) by creating a chimeric transcript using a splice acceptor sequence, or (4) by a combination of both an IRES and splice acceptor. Depending on the strategy, insertion of the reporter gene sequence can be within protein coding or untranslated exons or within introns. A conservative approach to obtain a faithful reporter is to make a simple insertion without deleting any sequence. This will avoid unintentional removal of regulatory elements in deleted sequence. However, you might consider a small deletion of the protein-coding sequences within an exon to increase the chances of generating a null allele. Introduction

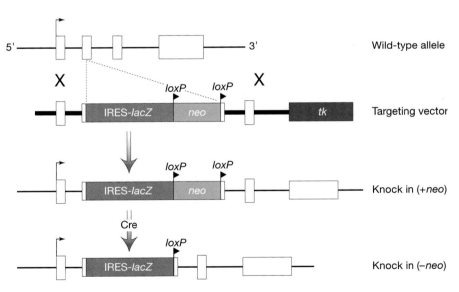

Figure 2.3. Reporter knock in. In this example, a *lacZ* reporter with a polyA is introduced into the second exon of a gene followed by a floxed *neo* expression cassette that can subsequently be removed either in targeted cells or in animals by Cre recombinase. An internal ribosome entry site (IRES) is used to bypass the requirement of having *lacZ* in frame with the coding region of the endogenous gene. An HSV thymidine kinase (*tk*) cassette is placed adjacent to the 3′ arm of homology for negative selection. The floxed *neo* cassette can be removed by Cre. Gene sequences can be deleted when making a knock in, but one must be aware that regulatory sequences might be removed. Instead of a reporter, the coding region of any gene of interest can be knocked in. Many times, knock ins also knock out endogenous gene function. (*Thick lines*) Regions of homology; (+*neo*) allele retaining the *neo* cassette; (–*neo*) allele in which the *neo* cassette has been deleted.

of a reporter into untranslated sequences could potentially yield wild-type, hypomorphic, or null alleles, depending on the specific situation.

Two notes of caution: (1) In spite of your best intentions and design efforts, knock ins will not necessarily reflect the endogenous expression pattern of the targeted gene. This is because the inserted reporter could act as a dominant insertional mutation by altering gene transcription. We just do not know enough about the regulation of genes to be able to avoid this in all cases. Thus, validation of the reporter knock in requires comparisons between endogenous and reporter expression. Once validated for specific stages and tissues, the knock in can be used with confidence to report the expression of the targeted locus during mutant analysis. (2) The activity of a reporter protein or its relative stability may result in a slightly different expression pattern compared with that of the endogenous mRNA or protein expression pattern. For example, X-gal staining reveals the enzymatic activity of the protein β-galactosidase, which may have different degradation kinetics from the endogenous protein of the mutated gene. If antibodies to the endogenous protein and the marker proteins are available, the extent of this discrepancy can be determined by comparing the two.

Knock-in reporters are useful for following the fate of mutant cells because they mark cells that transcribe the targeted locus. This can be very helpful for learning whether the expressing cells are still present in a mutant animal and, if so, how their behavior differs from wild-type cells. In addition, knock-in reporters can be used in combination with null alleles to determine whether the locus is regulated by negative feedback mechanisms. Assuming that the knock-in allele is also a null allele, the expression of the reporter in knock-in/wild-type heterozygotes is compared with expression in knock-in/null heterozygotes, both of which have a single knock-in reporter allele. If expression of the reporter is more extensive in the knock-in/null heterozygotes, this would suggest a negative feedback regulation of the gene by its own gene product.

Knock-in alleles are usually generated using a replacement gene-targeting strategy. The reporter is usually followed directly by a floxed positive selection marker gene, such as *neo*, that can be removed later by Cre recombinase. The 5′ and 3′ arms of homology flank the reporter-*neo* sequences and a negative selection marker can also be added. Strategies to identify homologous recombinants by Southern analysis are the same as for generating a standard null allele.

2.B.3 Producing point mutations or small changes

Another popular and versatile type of mutation to engineer is a point mutation (Figure 2.4). These can be used to create a stop codon that will truncate a protein product, alter transcription factor binding sites for transcriptional regulation studies, alter a specific amino acid by a missense mutation to create a variant protein, or perhaps mimic a mutation identified by human genetic studies. Many other types of alterations, such as splice mutations and untranslated region (UTR) modifications, are also possible with point mutations or other types of small changes.

There are three common methods for creating point mutations or other small changes by gene targeting in ES cells, each with advantages and potential disadvantages. These include (1) the Cre/*loxP* strategy, (2) the double-replacement strategy, and (3) the hit-and-run strategy.

1. *The Cre/loxP strategy uses a modified replacement vector design* (Figure 2.4A). The desired point mutation is placed within one of the two arms of homology so that after homologous

recombination, the point mutation is incorporated into the targeted locus. A floxed *neo*-selectable marker is usually placed within an intron at a site that will not influence the transcriptional regulation of the gene. However, there may be no prior information about transcriptional enhancers within introns, and so one way to stack the odds in favor of placing the floxed *neo* marker in a site lacking regulatory elements is to compare the intronic sequences of the gene between mouse and human. Conserved sequences suggest that they may have function and should therefore be avoided. Positive-negative selection is identical for a null allele, as are the type of probes used for Southern analysis. Ultimately, the *neo* marker is removed by Cre expression either in vitro or in vivo. This leaves a single *loxP* site in an intron but because this site could formally disrupt function, it is necessary to test a control allele in which the wild-type locus has a *loxP* site in the same position in the intron but lacks the engineered point mutation. This control can be generated either with an independent gene-targeting vector or from the same gene-targeting vector used to generate the point mutation (i.e., two mutations are independently generated from one targeting vector). This can be accomplished by having at least 1 kb of homology between the desired point mutation and the floxed *neo* marker. Thus, depending on where recombination takes place within the region of homology, one can obtain a targeted allele with or without the desired point mutation.

2. *The double-replacement strategy is a two-step process that employs two different replacement vectors* (Figure 2.4B). In the first gene-targeting step, a region containing the sequence into which you would like to introduce a point mutation is deleted and replaced with a *neo-tk* marker, using positive selection to identify the targeting event in ES cell clones. Once targeted clones have been identified, they are subjected to a second round of gene targeting using a replacement vector that spans the deleted region but contains the desired point mutation and no selectable marker. ES cell clones are selected for the loss of *tk* and screened for incorporation of the point mutation. In this strategy, no exogenous sequences (i.e., selectable markers or *loxP* sites) remain at the mutated locus.

3. *The hit-and-run strategy requires one round of gene targeting using an insertion-type vector that introduces the desired point mutation while simultaneously creating a duplication of all or part of the targeted locus* (Figure 2.4C). The targeting vector contains *neo*- and HSV *tk*-selectable markers. The targeting event is identified by positive selection of electroporated ES cells. Once targeted clones are identified, the cells of the clone are plated at clonal density and placed under negative selection. Clones that survive will have lost *tk* through intrachromosomal recombination and loss of the duplicated region, but the point mutation will be retained. Like double replacement, the advantage of this strategy is that no exogenous sequences remain at the targeted locus. This strategy is less often used because most investigators are less familiar with insertion vector strategies than they are with replacement vector strategies.

Regardless of the strategy that you use, one helpful tip to generate a point mutation is to engineer a restriction site by silent mutation for the desired amino acid change. This will allow you to genotype the point mutation more easily.

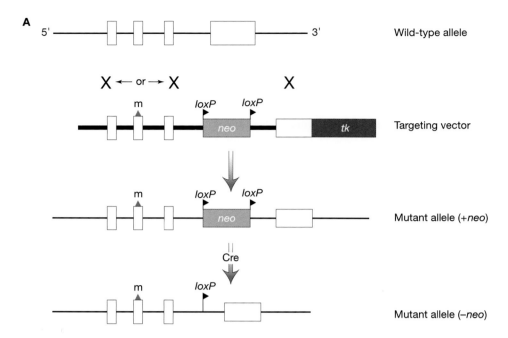

Figure 2.4. Engineering point mutations or other small changes. (*A*) Introduction of a point mutation using Cre/*loxP* gene targeting. In this strategy, a point mutation (m) is introduced into exon 2 using a standard replacement vector. The mutation is made in one of the arms of homology. A floxed *neo* expression cassette is placed within an intron and an HSV *tk* cassette is placed adjacent to the 3′ arm of homology. Two types of homologous recombinants can be obtained. One will have a crossover event 5′ of the point mutation, thus retaining the point mutation in the homologous recombinant (shown). The other will have a crossover event 3′ of the point mutation and 5′ of the *neo* cassette, leading to a targeted allele that does not have the point mutation (not shown). The floxed *neo* cassette is removed with Cre, resulting in a mutant allele with an intronic *loxP* site. (*Thick lines*) Regions of homology; (+*neo*) allele retaining the *neo* cassette; (–*neo*) allele in which the *neo* cassette has been deleted. (*B*) Introduction of a point mutation using double replacement. In this strategy, a point mutation (m) is introduced into exon 2 using a two-step procedure. In the first step, the region containing the exon you wish to mutate is deleted and replaced with *neo* and HSV *tk* expression cassettes using a replacment vector (targeting vector 1). The transfected cells are selected only for G418 resistance. Once homologous recombinants are identified, the ES cell clones are electroporated using a targeting vector (targeting vector 2) that spans the deleted region to bring in the desired point mutation. Loss of HSV *tk* is used for selection (i.e., correct targeting events will survive in ganciclovir or FIAU). The double-replacement procedure introduces the intended point mutation without leaving other sequences (e.g., *loxP* sites). (*Thick lines*) Regions of homology. (*C*) Introduction of a point mutation using the hit-and-run strategy. In this strategy, a point mutation (m) is introduced into exon 2 using a two-step procedure. In the first step, an insertion vector is created that contains the engineered point mutation in exon 2. After homologous recombination, the targeted locus has a partial duplication in addition to *neo* and HSV *tk* expression cassettes. One situation is shown in which the point mutation ends up in the 3′ duplicated region. However, it is possible that the point mutation could end up in the 5′ duplicated region, both the 5′ and 3′ duplicated regions, or neither. Partial tandem duplication results in a low frequency of spontaneous intrachromosomal recombination (this is the second step) that deletes one of the duplicated regions and the intervening sequence, resulting in a loss of HSV *tk*. This can be selected for by survival in ganciclovir or FIAU. The final ES cell clones must be screened to determine whether the point mutation has been retained (shown) or lost. The hit-and-run strategy introduces the intended point mutation without leaving any extraneous DNA sequences. (*Thick lines*) Regions of homology. (*Figure continued on facing page.*)

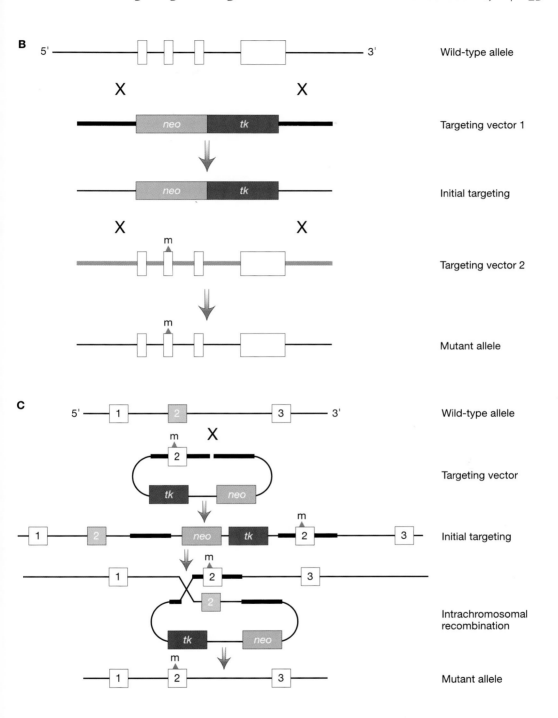

Figures 2.4B and C. (*See facing page for legend.*)

2.B.4 Conditional null alleles

Many genes are expressed during embryogenesis and in adult tissues and are therefore likely to have multiple roles in different tissues at different stages of development. Standard targeted mutations (e.g., null alleles) may cause early embryonic lethality, precluding the study of gene function at later stages of development or after birth. Conditional genetic strategies provide methods to bypass early lethality to study complex gene function. If you anticipate that your gene might fall into this category, and especially if you are mainly interested in the role of the gene in a specific tissue or organ, consider making a conditional null allele right from the beginning (Figure 2.5). See Chapter 8, Section 8.C to see what is involved.

Conditional null alleles are usually generated by targeting two *loxP* sites in the same orientation into noncoding regions (e.g., into introns or gene flanking regions) that flank critical protein-coding exons (Figure 2.5A). You may then use Cre recombinase at the time that you choose to act on these *loxP* sites to cause the deletion of the floxed exons. There are many gene-targeting strategies to make conditional alleles, but all require the use of a positive selectable marker, which may either remain in the conditional allele or be removed, for example, by

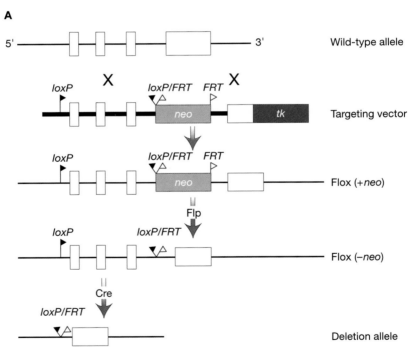

Figure 2.5A. Generation of a conditional null allele. In this strategy, a replacement vector is generated in which an *FRT*-flanked *neo* expression cassette is introduced into an intron. *loxP* sites are engineered to flank exons. It is very useful to engineer a restriction site 5′ of the 5′ *loxP* site to identify targeting events by Southern analysis that have included both *loxP* sites. An HSV *tk* cassette is placed adjacent to the 3′ arm of homology. After homologous recombination, a conditional allele is generated that has *neo* in an intron. This allele may behave as wild type, a hypomorph, or even a null. *neo* can be removed by Flp expression, yielding a conditional allele that will most likely behave as wild type unless the *loxP* sites have fortuitously disrupted an important regulatory sequence. Cre expression will delete the intervening exons to generate a loss-of-function allele. (*Thick lines*) Regions of homology; (Flox [+*neo*]) floxed allele that contains the *neo* cassette; (Flox [−*neo*]) floxed allele without the *neo* cassette.

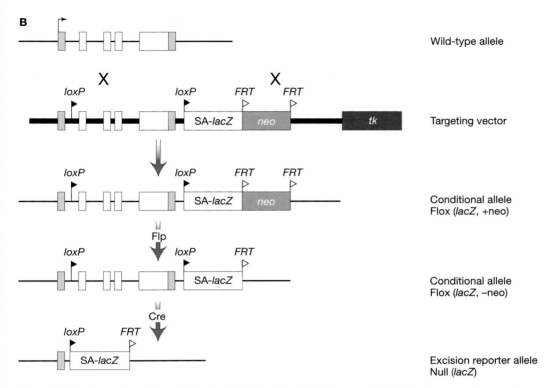

Figure 2.5B. Generation of a conditional null allele with a linked reporter. In this strategy, a replacement vector is generated in which a splice acceptor (SA)-*lacZ* pA cassette followed by an *FRT*-flanked *neo* expression cassette is introduced 3′ of the gene of interest. *loxP* sites are engineered to flank exons. It is very useful to engineer a restriction site 5′ of the 5′ *loxP* site to identify targeting events by Southern analysis that have included both *loxP* sites. An HSV *tk* cassette has been placed adjacent to the 3′ arm of homology. After homologous recombination, a conditional allele is generated that has the *lacZ* and *neo* expression cassettes 3′ of the locus. This allele may behave as wild type, a hypomorph, or even a null. The *lacZ* reporter should be silent because it is not transcribed and does not have a basal promoter. *neo* can be removed by Flp expression in vitro or in vivo, yielding a conditional null allele with a silent reporter. Cre expression will delete the intervening exons to generate a loss-of-function allele and a chimeric transcript containing exon 1 and *lacZ*, leading to β-galactosidase expression. (*Boxes*) Exons; (*shaded boxes*) untranslated regions; (Flox [*lacZ*, +*neo*]) floxed allele that contains the SA-*lacZ* and +*neo* cassettes; (Flox [*lacZ*, –*neo*]) floxed allele that contains the SA-*lacZ* cassette without the *neo* cassette; (Null [*lacZ*]) recombined allele with the SA-*lacZ* cassette.

using a second type of recombinase system such as Flp recombinase or ΦC31 integrase. In many cases, a conditional allele with the selectable marker in place can function as a wild-type allele. However, this will have to be determined for each individual case because the presence of the selectable marker and its promoter could potentially generate a null or hypomorphic allele. If the presence of the selectable marker results in a decrease in the level of gene expression, this hypomorphic allele could provide a useful genetic alteration for further analysis.

For simplicity, remove the selectable marker cassette either in vitro in ES cells before making chimeras (Box 2.3) or in vivo once you have mutant mice by crosses with recombinase-expressing mice (Section 4.D.1). Mice with the conditional allele should be phenotypically identical to wild-type mice and thus easily maintained as homozygous stocks. If this is not the case, you have a hypomorphic or null allele. Once you have a conditional allele in mice, introduction of Cre recombinase by breeding with *cre*-transgenic mice will result in Cre-mediated excision of the floxed coding exons, producing a deletion allele at a time and place of your choosing. For breeding strategies to make use of a conditional allele for time- or tissue-specific gene ablation, see Section 8.C.

> **HELPFUL HINT**
>
> *To be sure that the recombinase sites will function as planned in the targeted allele, it is worth testing for excision of the floxed or flrted exons in vitro in ES cells before mice are made. If you are removing the selection cassette in vitro (Box 2.3), you will recover clones with the exons removed as a matter of course. If you are waiting until later to remove the selection cassette in vivo by breeding, take the time now to express the appropriate recombinase(s) in vitro to ensure that a null allele can be produced.*

A conditional null allele with a linked reporter can also be engineered to monitor the excision of floxed exons by Cre recombinase using a reporter gene such as *lacZ* or *GFP* (Figure 2.5B). This type of allele can be very useful because it allows you to visualize recombination events (i.e., knockout of the gene of interest) with cellular resolution. This type of allele can be particularly useful if you do not have an antibody to the protein encoded by the gene of interest or if the cells that express the gene are scattered throughout organs, making RNA in situ analysis at the cellular level difficult.

There are a number of ways to design a conditional allele with a linked reporter. Basically, critical exons are flanked by *loxP* sites, a promoterless reporter gene is inserted downstream of the coding region of the gene of interest, and a selectable marker that subsequently can be excised in vitro or in vivo (e.g., using Flp/*FRT*; see Section 2.A.5) is incorporated for gene targeting. Ideally, the conditional allele should behave like a wild-type allele and not express the reporter. After Cre-mediated excision of the floxed exons, the allele should become null and express the reporter under the control of the endogenous regulatory sequences. In contrast to the "standard" conditional alleles in which only *loxP* sites are introduced into a locus, the likelihood of an introduced reporter altering the transcriptional regulation of the gene in question appears to be relatively high. In addition, just like any other knock in, the introduced reporter may or may not mimic the expression of the endogenous locus.

It is thus important to characterize thoroughly conditional alleles that utilize reporters, assaying for normal gene activity before Cre-mediated excision and validating the reporter expression pattern after Cre excision. To validate the reporter expression pattern, generate mice heterozygous for the recombined allele and compare the reporter expression pattern with that of the endogenous gene during development. This information will define the potential limitations of your reporter. In a number of instances, this type of conditional allele with a linked reporter of excision has acted as a hypomorphic allele before Cre-mediated excision (e.g., mice homozygous for the conditional allele or mice heterozygous for the conditional allele and a null allele of the gene show a phenotype). This would limit the usefulness of the conditional allele, but many experiments are still possible depending on the stage that you want to analyze and when the mice are compromised. The utility of this type of allele in contrast to standard conditional alleles is that recombination can be measured directly and with cellular resolution. So consider your experiments and decide on the type of allele that best suits your needs.

Now, go back to *gene-targeting Step 2* in Chapter 1, Section 1.A for the next step in using your gene-targeting construct.

CHAPTER 3

Getting to a Phenotype

THE MEANS TO AN END IN SCIENCE ARE OFTEN circuitous as unexpected results open new avenues of investigation. Sometimes this leads to serendipitous discoveries, but just as often it can be an annoyance and distraction if it delays the attainment of the intended end point. This chapter is designed to guide you through the minefield of technical and biological issues that can sidetrack the recognition of a mutant phenotype in a gene-targeting experiment, especially when the phenotype is not exactly as predicted. We assume that you have a source of proven germ-line competent embryonic stem (ES) cells and that you are following the detailed protocols provided by other sources [(MM3)* and see Appendix 1] for their culture, electroporation, and selection. The dichotomous branching format of this chapter provides you with the most efficient way of determining the effects of a mutation and steers you clear of time-consuming technical discoveries that others have made many times before you (see Box 3.1).

3.A **Gene-Targeting Step 3.** *Isolate clonal lines of correctly targeted ES cells and grow up a sufficient number for further experiments, freezing some for future use.*

Did you recover correctly targeted ES cells and grow up a sufficient number (see Box 3.2) for further experiments?

YES	Go to Section 3.B ▶
NO	See Possible Causes ▼

---

*MM3 refers to *Manipulating the mouse embryo: A laboratory manual*, 3rd edition (Nagy et al. [2003]. Cold Spring Harbor Laboratory Press, Cold Spring Harbor, New York).
Chapter opening artwork adapted from Chapman et al., *Mech. Dev.* **120:** 837–847 (2003).

> **Box 3.1** Ten Most Common Mistakes in a Gene-Targeting Experiment
>
> 1. Starting with "bad" ES cells, i.e., cells that will never give rise to germ-line chimeras.
> 2. Forgetting to add targeting vector DNA to the cuvette before electroporation; all ES cells die during selection.
> 3. Using wrong G418 concentration; none or all ES cells die.
> 4. Using β-mercaptoethanol at too high a concentration; all cells die.
> 5. Electroporating ES cells that were already drug resistant; all ES cells live.
> 6. Switching the lids of the 96-well plates during the freeze down of the master plate or inverting the orientation of the duplicate 96-well plate used for genotyping; using wrong ES cell clones to generate chimeras.
> 7. Killing ES cells during freeze down of the master plate; targeted clones can be identified but not retrieved.
> 8. Using wrong probe to identify homologous recombinants; no targeted clones can be identified or nontargeted clones were used to make chimeras.
> 9. Retaining herpes simplex virus thymidine kinase in ES cells, causing male chimera sterility; no germ-line transmission of the ES-derived genotypes (see Box 2.2, p. 16).
> 10. *loxP* site in the targeting vector is not functional; not discovered until after germ-line transmission and crosses with Cre mice.

Possible Causes *(in approximate order of decreasing likelihood)*

a. *Technical difficulties.* All ES cells die. If no ES cell colonies survived drug selection, it is worth considering that you may have forgotten to add the DNA targeting construct to the cell suspension before electroporation or that you added too high a concentration of the selective drug. Because the feeder cells are drug resistant, whether or not they die can provide an indication. Other technical problems that might lead to the death of all cells, such as problems with the electroporator, culture medium, incubators, etc., can easily be ruled out by the simple control of culturing two extra dishes of ES cells—one electroporated, the other not—without the selective drug. These ES cells should grow well, whereas the drug-treated cells should all die. Try another electroporation with this control, making sure that you add the DNA construct, to determine whether your electroporation and culture conditions are optimal.

b. *Further technical difficulties.* All cells survive. If a large number of clones or a solid lawn of ES cells survives, but few or none have incorporated the targeting construct, you may have forgotten to add the selective drug or used an inadequate dose of the drug. Or perhaps you thawed the wrong vial of ES cells for the electroporation and used some that were already drug resistant. It is always worth determining a "kill curve" for a particular ES cell line and batch of drug so that you know what to expect. Simply determine a dose-response curve for the drug, selecting the lowest dose that kills 100% of the cells within 5–7 days.

c. *Trouble with the targeting construct.* If you are getting drug-resistant colonies with random integrants of your targeting construct, but no homologous recombination at the targeted locus,

> **Box 3.2** **Necessary and Sufficient Number of Targeted ES Cells**
>
> How many clones are enough?
>
> To verify that a phenotype is the result of the targeted mutation you created, it is useful to show that two independently targeted ES cell clones give the same phenotype. However, to be on the safe side, it is worth recovering more than two targeted clones: five to ten would be reasonable. These cells have been subjected to electroporation, drug treatment, and cloning. Even if you started with a parent ES cell line with excellent germ-line potential (which you should have tested), the resultant clones could have picked up mutations, chromosomal aberrations, or restrictions of developmental potential during these procedures, so that any single clone might not provide germ-line transmission through chimeras. Similarly, if you have only one or two clones, they might succumb to some technical glitches along the way, so it is wise to have a few more as backup.
>
> In the event that your targeting efficiency is very low and it is hard to get a reasonable number of targeted clones without massive effort, check the technical aspects of the electroportation procedures since this is often where the problem lies. Check the DNA concentration and purity. Make sure that the DNA is completely dissolved and well mixed before taking an aliquot for electroporation. In addition, make sure to use the correct cell concentration and electroporation parameters. If the efficiency is still low and you have consulted Chapter 2 for hints on optimizing the targeting construct, persevere a bit longer. In our experience, a definite learning curve exists and gene-targeting efficiency often improves with electroporation practice.
>
> How many cells are enough?
>
> As soon as you have identified a targeted clone, grow it up, verify that it is indeed targeted by repeating a Southern analysis, and freeze down a sufficient number of cells for future experiments. Five to ten vials should be fine, but you might want to place them in several different locations to guard against losing them all in a disaster.

it could be that you have a less-than-perfect targeting construct or that your locus is hard to target. (Before going further, make sure that you are using the correct genotyping probe. Is the band of the correct, predicted size? It is worthwhile to sequence the probe fragment to ensure that it is correct.) Some loci are notoriously difficult to target but we have yet to hear a convincing explanation for this, or more importantly, how to predict it. Your best bet is to review Chapter 2 and make sure that you have done everything you can to optimize the chances of homologous recombination, particularly with respect to the use of isogenic DNA and to having sufficiently long arms of homology. If you have any doubt, consider redesigning and rebuilding the targeting construct. However, before you do, electroporate again with the construct that you have and aim for a large number of clones. You may have a low targeting efficiency that could require screening 1000 clones or more, but this usually takes less time and effort than rebuilding the targeting construct. On the other hand, the low efficiency may be part of a learning curve that will improve with practice, so do not get discouraged.

d. *Inactive selectable marker.* All ES cells die, but you have ruled out the technical difficulties listed in "*a*" above and drug-resistant feeders are fine. Although it is a fairly remote possibility, the selectable marker in the construct may be inactive, either because of mutations introduced during cloning or the activity of a transcriptional repressor in the region of homology used for building the construct. In the first case, you could sequence the selectable marker you used and replace it with another if it has been mutated. If this is not the problem, consider a different placement of the selectable marker in the construct, e.g., use a different region of homology in hopes of avoiding the putative repressor.

e. *PHENOTYPE!* Finally, there is the possibility that *this is your phenotype*! Perhaps the mutation of a single copy of your favorite gene is sufficient to render the heterozygous ES cells incapable of growing—a dominant effect. Explore this last possibility but continue to work, because we know of no actual case of a gene with this particular heterozygous mutant effect of killing ES cells. But how would you go about testing it if you think that you have that rare case? Go to Chapter 7. If you can rule out this possibility, try electroporating again and select a large number of clones for further analysis. You simply may not have tested enough clones to recover a rare homologous recombination event. If all else fails, make a new targeting vector incorporating greater or different regions of homology.

HELPFUL HINT

X-linked, Y-linked, and imprinted genes are special cases in that only one copy is present or active, respectively, so the cells are functionally hemizygous for the gene. The possibility that a mutation in the single copy or single active copy results in a phenotype is thus more likely since it does not require a dominant effect.

3.B Gene-Targeting Step 4. *Produce ES cell chimeras by injecting the correctly targeted ES cells into mouse blastocysts or by doing morula aggregation (MM3).*

Ideally, the ES cells will make a large contribution to the chimera that will be detectable by a high contribution to the coat color phenotype (i.e., high-level chimeras). However, there are many other possibilities.

Did you get chimeras with your targeted ES cell clone?

YES	Go to Section 3.C ▶
NO	See Possible Causes ▼

3.B.1 No mice born at all

Possible Causes

a. *Technique, technique.* Practice control embryo transfers until you have a good success rate. You should be able to achieve an 80–90% success rate in embryo recovery after embryo transfer before you start with important experimental embryos. Otherwise, you are wasting your time. Try to reduce surgical trauma and, in particular, treat the uterus gently, because rough treatment can cause a pregnancy to fail. Make sure your embryo culture medium is correct (MM3). In addition, be sure that your pseudopregnant embryo transfer recipients are

Getting to a Phenotype | 33

> **Box 3.3** **Determining the Number of Oocytes Ovulated by Counting Corpora Lutea**
>
> Corpora lutea (CL) are the visible and lasting signs of ovulation. At the sites of extrusion of oocytes from the surface of the ovary, the ovarian follicles form endocrine tissues known first as corpora hemorrhagica and later as CL. These structures are essential for the maintenance of pregnancy since they are a major source of progesterone. They are also useful diagnostic aids for two specific experimental situations. One is during embryo transfer to pseudopregnant recipients. If there are no CL, the recipient did not ovulate and is therefore not pseudopregnant; she should not be used for embryo transfer. The second situation is in the diagnosis of preimplantation embryonic phenotypes, when the CL can be taken as a measure of the maximum number of embryos that could be expected, provided all ovulated oocytes were fertilized.
>
> To count CL, remove the ovaries and place them in medium or phosphate-buffered saline (PBS) in separate drops marked right and left, so that you can eventually correlate the counts with embryo counts from each oviduct or uterine horn. Under a dissecting microscope with mostly top lighting—not the transmitted light you use for preimplantation embryo collection—look for uniform, raised, pink hemispheres (approximately 0.75 mm in diameter) on the surface of the ovary. Turn the ovary over to ensure that you find them all. The CL are generally rosy pink compared to the whiter background ovarian tissue (Figure 3.1). If you use a little transmitted light as well, they appear more dense and uniform compared to the relatively transluscent, knobbly remainder of the ovary. Ideally, for unbiased data, it is best to do these counts blind with respect to the number of embryos actually recovered.

really pseudopregnant (male mice have been known to plug females that are not in estrus) by checking that they have corporea lutea (CL) in the ovary at the time that you do the embryo transfer (Box 3.3). If not, you are essentially throwing the embryos away, so use another recipient. Finally, check the light cycle control in your animal room to ensure that it is turning on and off correctly (14 hours light/10 hours dark is good, but 12/12 is fine, too) since erratic lighting disturbs the reproductive cycle.

b. *Maternal issues.* It is possible, especially if you saw that the recipients were pregnant (i.e., they looked bulgy around the middle 10–12 days after embryo transfer), that pups were born but cannibalized before you had the chance to see them. Be aware of the date that the chimeras are due to be born and search the bedding material for telltale remains on that date

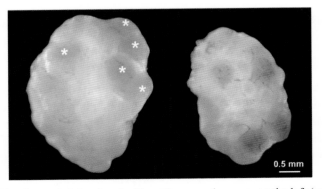

Figure 3.1. Two ovaries from a 2.5-dpc female. Five CL are on the ovary on the left (*marked with asterisks*) and five are on the ovary on the right that are not marked. Can you find them?

(if you find any, even small remains can be genotyped to determine whether the newborns were chimeric). Examine the uterus of the mother shortly after the expected due date for signs that she recently gave birth. There will be red parturition sites on the mesometrial side of the flaccid, distended uterus if she gave birth.

If cannibalism seems to be the problem, there are several ways to improve this antisocial maternal behavior. Minimize crowding during late pregnancy: Keep only one or two pregnant mice in the same cage (however, note that local institutional rules may dictate the number of animals per cage). Avoid a bedding or cage change within 1–2 days of the expected birth date, but be sure that the cage is relatively clean. This means that you may have to put a "do not disturb" sign on the cage and take the responsibility of keeping it clean, fed, and watered rather than relying on animal husbandry staff. Provide nesting material a few days in advance of the due date. You can use commercial nesting materials or you can place a tissue or paper towel in the cage. No matter how anxious you are to see the babies, do not disturb the mother too much during or shortly after birth. If you observe the pups strewn around the cage and/or mutilated, then and only then is the time for drastic intervention in the form of cross fostering onto a lactating female (see Box 3.4). If cannibalization is a persistent problem, a last resort is to preempt the birth by delivering the pups by Caesarean section (MM3) a day before the expected birth date and cross fostering onto a lactating female.

c. *Phenotype.* It is just possible that the mutant ES cells have caused the death of all chimeric embryos if the mutation is dominant and has a very detrimental effect or if the gene is X linked or imprinted. Looking at the uteri of the foster mothers shortly after the expected birth date will indicate whether any postimplantation embryonic development took place, since the implantation sites or resorption sites (places at which embryos implanted but then died and were partially resorbed by the uterus) should still be visible as swellings or blood spots on the uterus. If you suspect that the chimeras are dying, either before or after implantation, try including in the embryo transfer a few wild-type embryos of a distinct coat color along with the chimeras to determine if death is specific to the chimeras. If all of the embryos fail, it is more likely to be the result of a more general problem as detailed in the previous sections. If still no mice are born even with the inclusion of nonchimeras in the mix, go back to Section 3.B.1.a and polish your technique. If nonchimeric embryos develop but the ES cell chimeras do not, you might just have a dominant effect. Go to Chapter 7.

3.B.2 Pups are born but none are chimeric or the level of chimerism is very low

Possible Causes

a. *Technique.* Were the ES cells kept under optimal conditions during the injection experiment? Trypsinized cells should be held on ice in buffered medium in a capped test tube. You can check if this manipulation is compatible with ES cell viability by culturing the "leftover" ES cells after a blastocyst injection or morula aggregation experiment under standard ES cell culture conditions to see if they still grow well. Some of these trypsinized cells are transferred into the injection or aggregation dish for chimera production. pH and osmolarity must be correctly maintained in the injection or aggregation dish. Watch the color of the medium and use a sufficient volume of medium in the dish to reduce concerns of evaporation. Are you injecting 12–15 ES cells, which seems to be the optimal number for most ES cell lines?

Box 3.4 Cross Fostering to Counteract Bad Parenting

Sometimes the maternal instinct is just not what it should be, especially in first-time mothers. A number of factors, such as overcrowding, disturbance, an alteration in the light cycle, temperature, or the introduction of a strange mouse—male or female—around the time of parturition, can lead to maternal neglect of offspring, or worse, cannibalization. If you catch the mother in the act, the pups might be scattered around the cage, cold to the touch, and possibly have bite marks. If the pups are particularly valuable, it is worth trying to cross foster them onto a lactating female. If they are breathing and not actively bleeding, you may be able to rescue them. This same method can also be used following the Caesarean delivery of offspring (MM3).

- Remove the abused pups from the cage and warm them up. *Do not* use a lamp, since this is likely to overheat and dehydrate them. Place the pups in a petri dish on a slide warmer set at 37°C or hold them in your hand until they are warm.
- Locate a lactating female who has given birth within the past day or two and is successfully nursing her young. She can be any strain, but determine how you will distinguish her pups from the foster pups later, possibly by incorporating a coat color marker (see Box 6.1, p. 139). If the foster mother's litter is large, decrease the size by removing as many of her pups as you plan to foster. Remove the biggest, healthiest ones to give the foster pups a fighting chance.
- Remove the foster mother to a holding cage while you try to fool her into thinking that these are all her babies. Take both sets of pups and some soiled bedding from the foster's home cage and mix well. The pups should be at the same body temperature. Just for good measure, we suggest adding a few drops of the foster mother's urine to the mix (simply hand restrain her above the babies and let it drip). We know of one investigator who briefly anesthetizes the foster mother supposedly to induce amnesia, but we have not found this to be necessary.
- When you believe the pups are ready, *active* and *warm*, and smelling of the foster family's home cage, place them in the nest and reintroduce the foster mother. You will know immediately if she is going to accept the babies. If not, they will be dumped outside the nest in rapid order.

Are you confident that your blastocyst injection or morula aggregation procedure is effective and that the ES cells are incorporated into the embryos before transfer to the foster mother? If you think that the lack of incorporation of ES cells might be a problem, you could try injecting or aggregating some marked ES cells, for example, cells with a *lacZ* or a green fluorescent protein reporter gene; culturing the chimeric embryos overnight; and then looking for the cell marker as an indication of ES cell contribution. If marked cells are not present, it is a good idea to reevaluate your injection or aggregation technique as well as your ES cell-handling technique, because the cells might have died or they might never have been incorporated. If the cells are present in good numbers and appear to be incorporated into the embryo, continue working through the other possibilities.

HELPFUL HINT

Always check that your parental ES cell line is competent to make chimeras and to contribute to the germ line before using it for gene targeting. You can do this yourself by making test chimeras with the parent ES cell line or you can depend on information supplied by the person who gave you the ES cells in the first place.

> **HELPFUL HINT**
>
> *One common reason for a restriction in germ-line potential is that the ES cells have become aneuploid. If this has happened, a subpopulation of euploid cells might still be in the culture and if the clone is a valuable one, i.e., you have only one targeted clone, it might be worth trying to isolate the euploid cells by subcloning and karyotyping cells (MM3) and then making chimeras with euploid subclones.*

b. *The targeted cell clone has lost developmental potential.* Try another clone. You should have more than one if you followed our recommendation above. If not, go back to *gene-targeting Step 3* and isolate more targeted ES cell clones.

c. *The host embryo strain may not be conducive to making chimeras with your ES cells because you did not use the standard combinations.* Most laboratories use 129-ES cells and C57BL/6 embryos for blastocyst injection because the 129 component has a competitive advantage over the C57BL/6 component in contributing to somatic and germ-line tissues in chimeras. Albino C57BL/6 strains have been developed and are also used for recipient blastocysts for injections. If chimeras are generated by morula aggregation, most investigators use albino or black outbred mice for the embryonic component because many embryos can be obtained much less expensively than C57BL/6 inbred embryos. Try again with larger numbers or consider changing to a different strain of mouse for the host embryos and starting again from *gene-targeting Step 4*, making more chimeras.

d. *Possible coat color phenotype.* If offspring are all nonchimeric by coat color, but the sex ratio is skewed heavily in favor of males, consider that the mutation you made may have affected the coat color. Maybe some of the offspring are actually chimeras, hence the skewed sex ratio, but the coat color is not as expected because of the heterozygous mutation. This is fairly unlikely, but it has been known to happen. The phenotypic sex of XX↔XY chimeras will be skewed in favor of males, assuming that you are using an XY ES cell, because of the nature of sex determination. Genotype some of the offspring to be sure that they are not cryptic chimeras. If they are, you will have an interesting dominant phenotype. Go to Chapter 7. Otherwise, make more chimeras with a different clone.

e. *Could this be a phenotype?* Consider the possibility that cells heterozygous for the mutation are incapable of contributing to chimeras because of haploinsufficiency for the gene product, if you made a null mutation, or because of a defective gene product, if your mutation results in a neomorphic or antimorphic protein. If the targeted ES cells grew well, the defect might only become evident as the ES cells differentiate (or try to) in the embryo in vivo. Go to Chapter 7.

> **HELPFUL HINT**
>
> *If the gene is X linked, the mutation will affect the only copy of the gene in an XY ES cell line, or if it is imprinted, the mutation could affect the only active allele depending on the direction of imprinting and the tissue in question.*

3.C Gene-Targeting Step 5. *Breed the male chimeras to recover offspring carrying the targeted mutation through germ-line transmission of the targeted allele (see Box 3.5).*

Do chimeras produce offspring derived from the ES cells?

YES	Go to Section 3.D (or Section 3.C.3 if only the wild-type allele is recovered in the offspring) ▶
NO	See Possible Causes ▼

3.C.1 The chimeras produce no offspring at all

Possible Causes

a. *Unexplained infertility.* This is a good, catchall phrase and could apply to either your chimeras or the mice that you have mated with the chimeras. Changing the females (selecting young, estrous females) (see Box 3.6) will provide the chimera with a change of scenery, but do not spend too much time here. It is only if you have very few chimeras that you should waste breeding space on nonproductive test-matings, but even then, go back to *gene-targeting Step 4* to make additional chimeras so as not to waste precious time. After all, the goal is to have a germ-line transmitter from more than one ES cell clone.

b. *Intersex chimeras.* The most common cause of a chimera producing no offspring is that the chimera is neither male nor female, i.e., an intersex. Most ES cells in common usage are XY (i.e., they were originally derived from male embryos) and you have a 50:50 chance that any injected blastocyst will be a female (XX). Usually, XX↔XY chimeras develop as phenotypic males and fortuitously, the biology is on the side of the investigator: This is highly desirable because only the XY (ES-derived) cells make functional sperm. However, possibly due to the level of contribution of the XY ES cells, XX↔XY chimeras sometimes have both male and female reproductive organs or ovotestes and consequently they are infertile. Examination of the gonads will tell you whether this is a possibility; intersex gonads have areas of ovarian and testicular tissue and the uterus, vas deferens, and/or external genitalia may be abnormal. Before sacrificing the chimera, you could do a laparotomy using anesthesia if you wanted to be sure that this is the cause of infertility. Even if a hermaphrodite produces copulation plugs, it is highly unlikely that it will be of any further use. In this case, go back to *gene-targeting Step 4* and make more chimeras.

c. *Dominant sex reversal.* It is possible that mutation of the gene in question could cause a dominant sex reversal even in chimeras. Because most investigators use XY ES cells, the consequence would be the generation of predominantly phenotypically female chimeras that still retain the Y chromosome and that would be unable to produce viable oocytes (or sperm). If all of your chimeras are female and infertile, you can determine whether they all have a Y chromosome using a Y-specific probe or PCR primer set (see Box 3.7), remembering that half of them should have a male component derived from the embryo. If it turns out that they all have a Y chromosome, go to Chapter 7 for the next step.

Box 3.5 Breeding Schemes for ES Cell Chimeras

The object of test-breeding ES cell chimeras is to identify the chimeras that transmit the ES cell genotype through their germ line (so-called germ-line transmitters) and to produce as many ES cell offspring as possible in the shortest time with the least amount of work. We take the common case of chimeras made by combining a 129-derived ES cell line with C57BL/6 embryos, but the principles are the same for other strain combinations.

The first decision is whether to breed female and male chimeras, or just the males. Because most ES cell lines are male derived (XY), the sex ratio of chimeras is skewed in favor of males because XX ↔ XY chimeras tend to develop as males. The infrequent female chimeras do not generally transmit the ES cell genotype because XY cells undergo meiosis in an ovary rarely, if at all, and it is thus not worth test-breeding them. However, if you notice that the sex ratio of chimeras is not skewed and that there is an equal number of high-level female chimeras, it is worth test-breeding them since this might indicate that the Y chromosome has been lost, and XO cells make perfectly good oocytes.

With what should you breed the chimeras? Several options follow:

- *Breed chimeras with C57BL/6 mice* (see Figure 3.4). With this combination, you can take advantage of the coat color difference between the strains: All pure black (nonagouti) offspring are from the C57BL/6 component of the chimera and all agouti (yellowish brown) offspring are from the ES cell component (agouti is dominant over nonagouti). It is only necessary to genotype the agouti offspring; they are distinguishable from the black ones by 7 days of age when the yellowish, agouti hairs begin to show at the back of the neck. Half of the agouti offspring should be heterozygous for the targeted mutation, and, as a bonus, they will all be otherwise genetically uniform (129 × C57BL/6)F_1 mice. Another advantage of this breeding scheme is that the black offspring, which are all pure C57BL/6, can be raised and recycled for use in test-breeding and/or additional chimera experiments.

- *Breed chimeras with 129 mice.* The main advantage of this scheme is that the resulting ES cell-derived offspring will be co-isogenic with the 129 strain; there is no need for lengthy backcrossing to have the mutation on an inbred background. The disadvantages are that 129 strain mice are not very good breeders, and you are giving up the possibility of using a coat color marker to distinguish ES-derived from C57BL/6-derived offspring of the chimera. Only agouti offspring will be produced and therefore all will have to be genotyped by PCR or Southern analysis. This scheme is most useful for producing a co-isogenic strain after you have identified a chimera that transmits only the ES cell genotype (a 100% germ-line transmitter) using the previous, C57BL/6 breeding scheme.

- *Breed with outbred mice.* The biggest advantage of this scheme is the increased fecundity usually associated with outbred mice. However, unless you choose the strain with care, you may be giving up the possibility of using a coat color marker to distinguish ES cell-derived offspring. Pick a strain, albino or not, that is known to be homozygous nonagouti (Black Swiss is one possibility). If you use an albino mouse, also be aware that some ES cell lines (notably, R1) segregate for albino alleles that might give you white or dilute coat colors—but at least you know where they came from. The main disadvantage (some might say advantage) of this breeding scheme is that your mutation will be on an outbred genetic background. Nevertheless, you can always switch to another breeding scheme once you have identified the germ-line transmitting chimeras if you want to have the mutation on a defined genetic background.

Box 3.6 Selecting Estrous Females

The estrous cycle in the mouse ranges from 3 to 4 days and ovulation occurs around the midpoint of the dark period of the light cycle. Mating also occurs around the middle of the dark period, but a lot of variation is possible. Many animal rooms are set to a 12 hours light/12 hours dark cycle, but a 14/10 light/dark cycle tends to make the time of ovulation and mating tighter and you might notice less variability in embryonic stages using this cycle. One way of getting timed pregnancies is simply to leave females with males continuously, but a more efficient method is to select estrous females and place only those in proestrus or metestrus with males. In either case, check for vaginal copulation plugs in the morning, because they can fall out or dissolve during the course of the day. These plugs are white or cream colored, solidified components of the ejaculate (Figure 3.2) and simply indicate that mating has taken place. Estrous females can be recognized by the appearance of the external vaginal epithelium. Lift the tail of the female and *gently* probe the vagina (the same way you would check for a vaginal plug the next morning), so as not to induce pseudopregnancy. You want to select females with two or more of the following vaginal characteristics:

- dry but not flaky;
- pink, not red, white, or blue;
- swollen, so that the tissue bulges outward;
- epithelium that folds into corrugations on upper and lower vaginal lips.

Figure 3.2. Appearance of a vaginal plug on the morning after copulation. The ejaculate from the male forms a solid plug in the vagina that can remain there for 8–24 hours. A probe, such as the dental spatula shown here, can be used to gently probe the vagina to detect the plug. (Reprinted, with permission, from Papaioannou and Johnson, *Gene targeting: A practical approach* [ed. A.L. Joyner], pp. 133–175, Oxford University Press, United Kingdom [2000].)

> **Box 3.7** Genotyping for the Presence of the Y Chromosome
>
> PCR is the simplest method to determine if your mouse carries the Y chromosome (XY) or not (XX or XO). Primers for *Sry, Zfy*, or other Y-linked genes can be used. Primers for *Sry* and an autosomal, internal control gene (*Rapsn*) are given below.
>
> The mouse Y chromosome carries many repeat sequences. One sequence, called Y353/B, recognizes Y-chromosome-specific repeats that are transcribed in the testis. It can be used as a probe for DNA spotted onto filters or on Southern blots.
>
> *Sry* PCR genotyping
>
> *Sry*-forward: 5′ TGACTGGGATGCAGTAGTTC 3′
> *Sry*-reverse: 5′ TGTGCTAGAGAGAAACCCTG 3′
> *Sry* fragment size ~230 bp
>
> *Rapsn*-forward: 5′ AGGACTGGGTGGCTTCCAACTCCCAGACAC 3′
> *Rapsn*-reverse: 5′ AGCTTCTCATTGCTGCGCGCCAGGTTCAGG 3′
> *Rapsn* fragment size ~590 bp
>
> Cycle conditions: 95°C, 5 min; 35 cycles of 95°C for 30 sec, 65°C for 30 sec, 72°C for 45 sec; 72°C, 10 min; store at 4°C.

3.C.2 Chimeras' offspring are from the blastocyst component but not the ES cells, i.e., there is no germ-line transmission

Possible Causes

a. *Impatience.* The absence of germ-line transmission of the ES cell component many simply indicate that you are overanxious and the sample is too small to draw conclusions. Whole litters may be derived from one component of the chimera or the other, possibly because of patches of cells populating a testis tubule, and the composition of litters can change with time. Remember that the number of offspring from any chimera gives you only a sampling of the germ line of that animal. If you have a chimera with a 5% contribution of ES cells to the germ line, then you can expect only one out of 20 offspring to be of the ES cell genotype; it is all about probability. Generally, the level of chimerism in the coat (assuming that you are using a coat color marker to assess chimerism) (see Box 3.8) is a guide to the contribution of ES cells to the germ line, but chimerism in the coat (Figure 3.3) is no guarantee of chimerism in the germ line (or any other tissue, for that matter). Here is where your gambling instincts can come in handy: If you have only low-level contribution of ES cells to chimeras, it may not be worth breeding them very extensively. If you have some chimeras with low-level contribution and some with high-level contribution, limited animal space is better used by concentrating on the game with better odds. In other words, breed only the high-level chimeras. If you have no success with further test-breeding, go back to *gene-targeting Step 4*, and make more chimeras. See Section 3.B.2 for pointers on improving the level of chimerism.

Box 3.8 Common Coat Color Markers

The rainbow-hued litters that emerge in the second generation of breeding a targeted mutation often surprise researchers. But it is not for nothing that coat color genetics has been the darling of the mouse fancy for hundreds of years. Do not be worried by the segregation of coat color genes and do not be fooled into thinking it is a phenotype. A look at coat color genes carried by ES cells and a quick review of dominance and epistasis are in order. The most commonly used ES cells were originally derived from 129 mice. However, the 129 "strain" is anything but uniform (see Box 4.3, p. 53); it has many different substrains with different combinations of coat color alleles among them. If the parental ES cell line used for gene targeting came from mice heterozygous for coat color alleles, these will eventually segregate in the offspring of a germ-line transmitter to provide you with the entertainment of trying to decipher complex coat colors. When you choose an inbred strain for test-matings, you know that the mice are homozygous at all loci, but cross these with a segregating strain and the rainbow unfurls (see Figure 3.4).

Dozens of loci affect coat color, but for practical purposes, only a few need to be considered. One is the *albino* locus. The most recessive allele, *albino* or *Tyr^c*, results in no pigment at all in the homozygous state and so is epistatic (takes precedence) over all other coat color alleles and essentially hides them (thus, by looking, you cannot tell the other coat color alleles that an albino mouse carries). Many random-bred mice are homozygous for *albino*; this genotype (*Tyr^c*/*Tyr^c*) results in the standard pink-eyed, white laboratory mouse—but they might be heterozygous for alleles at other coat color loci and some suppliers indicate known segregating coat color loci in their catalogs. There are also other alleles at the *albino* locus, such as *chinchilla* (*Tyr^{c-ch}*/*Tyr^{c-ch}*), which reduces but does not eliminate pigment; thus, the result is a dilution of whatever other colors are present. The "wild type," *Tyr$^+$*, is non-albino, resulting in full pigmentation in coat, skin, and eyes. It is fully dominant to all other albino alleles.

A second locus of interest is the *agouti* locus, which affects the distribution of pigment types, yellow or black, throughout the hair and coat. A large allelic series can be found at this locus, with the bottom recessive allele resulting in exclusively black pigment and the top dominant allele resulting in exclusively yellow pigment (in the hair). The most common alleles, *agouti* (*A*) and *white-bellied agouti* (*A^w*), result in banding of individual hairs with yellow and black pigments and a belly lighter than the back. These alleles are commonly referred to as "wild type," because variations of these alleles occur in nature in many mammals and result in a brownish, mousy color that is an excellent protective camouflage. This is the "agouti" mouse, whereas the "nonagouti" mouse is typically solid black.

Common ES cell lines, such as R1, D3, and AB-1, were derived from 129 mice homozygous for *white-bellied agouti*, or *agouti* alleles at the *agouti* locus. Some of these, however, were heterozygous for *albino* and/or *chinchilla* and also for different alleles at another dilution locus called *pink-eyed dilution* (*p*). The interplay of these alleles with the alleles carried by the strain to which you mate your germ-line chimera will result in white, dilute, solid color, and agouti mice with black or pink eyes. See Figures 3.3 and 3.4, and if you want more information, refer to the sources in Appendix 1.

HELPFUL HINT

If you have used the common combination of 129-ES cells and C57BL/6 blastocysts, you can use your test-breeding program as a production colony for C57BL/6 mice by test-mating chimeras with C57BL/6 mice. Germ-line transmission is detectable by an agouti phenotype in the offspring at 7 days of age. The black offspring are pure C57BL/6 mice and can be recycled straight into your breeding or production colony, saving purchase costs.

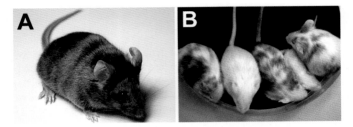

Figure 3.3. Examples of chimera coat colors. (*A*) The typical case of a 129-derived ES cell line combined with C57BL/6 (B6) blastocysts. The chimera has patches of agouti and black hairs by virtue of the *white-bellied agouti* (A^W/A^W) genotype of the ES cells and the nonagouti genotype (*a/a*) of the B6 embryos. This animal is somewhat greater than 50% agouti and would be a reasonable candidate for test-breeding. (*B*) Chimeras made with 129-ES cells and albino, nonagouti blastocysts (Tyr^c/Tyr^c; *a/a*) have white, black, and agouti patches. The white patches are areas populated by blastocyst-derived albino melanocytes, the agouti patches have 129-derived melanocytes and hair follicles (the site of action of the *agouti* gene), and the black patches have 129-derived melanocytes but blastocyst-derived (*a/a*) follicles. These animals range from about 5% to 50% chimerism.

b. *Unintended consequences of targeting construct design.* By any chance, is there a *tk* gene in the targeted allele? An active herpes simplex virus thymidine kinase (HSV *tk*) gene is often used for negative selection in gene-targeting experiments. However, HSV *tk* expression is incompatible with spermatogenesis. If you used HSV *tk* in your construct, you must remove it before making chimeras (see Box 2.2, p. 16). Either go back to *gene-targeting Step 1* and remake the targeting construct by putting in a different selection cassette or, if the HSV*tk* has *loxP* or *FRT* sites around it, go back to the targeted ES cells and remove HSV *tk* by transient transfection with the appropriate recombinase (see Box 2.3, p. 17) and then start again from *gene-targeting Step 3*. In addition, an HSV *tk* gene that has been modified such that it does not cause male sterility can still be used as a selectable marker. If you have a strong reason for needing HSV *tk* in mice, investigate this altered gene.

c. *Lost potential.* The ES cell clone may have lost the potential to contribute to the germ line and proceed through spermatogenesis (or oogenesis, if you are using an XX ES cell line) through some effect unrelated to the targeted allele. Although it will be hard to prove that this is the cause, and it is probably not worth the effort, you might reach this conclusion if you have produced a large number of high-level, breeding chimeras, none of which transmits the ES cell genotype. This is a signal to move on to another targeted clone. If the same happens with multiple targeted clones, karyotype the clones and the parent cell line to determine if there is a high level of aneuploidy. If not, perhaps you have a phenotype; see the next paragraph.

d. *Phenotype alert!* It is possible that the specific mutation you made renders the heterozygous cells incapable of contributing to the germ line or completing gametogenesis, even though they make cellular contributions to other tissues in chimeras. This might apply to only a very small number of genes. But if you think this is possible based on the expression pattern of the targeted gene during gametogenesis or on the nature of the protein, for example, as a last resort, go to Chapter 7 to investigate this possible dominant effect.

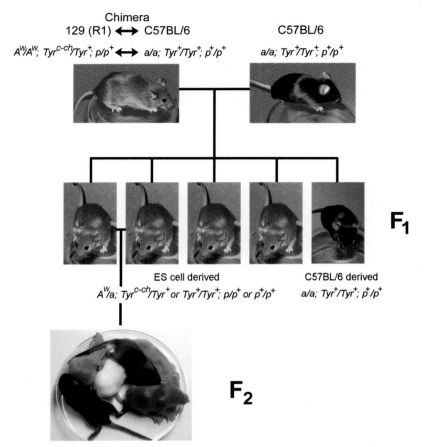

Figure 3.4. Example of coat colors obtained in a typical chimera test-breeding program. In this case, a near 100% chimera made with R1 129-ES cells (indistinguishable from a wild-type agouti mouse) and a C57BL/6 (B6) blastocyst is crossed with a B6 mouse. All of the F_1 offspring are agouti or black, depending on whether they come from the ES cell-derived or B6-derived germ cells of the chimera, respectively. When heterozygous ES cell-derived offspring are intercrossed, segregating recessive alleles at the linked *albino* (*Tyr*) and *pink-eyed dilution* (*p*) loci from the R1 ES cells, as well as the segregating agouti alleles, provide a spectrum of coat colors in the F_2 generation ranging from agouti and black through various shades of agouti or nonagouti dilute color. These coat color alleles will segregate independently from the mutant allele, unless, of course, the mutant allele is linked to one of the coat color loci. In the case of ES cell lines, such as AB-1, which are not segregating for *albino* or *pink-eyed dilution* alleles, offspring of F_2 mice will be either black or agouti, independent of the mutant allele.

3.C.3 Offspring are produced from the ES cell component, but only the wild-type allele is recovered

HELPFUL HINT

Take heart. It can be very unsettling when chimeras fail to transmit the targeted allele. But rest assured that if your chimeras transmit the ES cell genotype, at least your parent ES cells and clones have good developmental potential and your blastocyst injection or embryo aggregation skills are good, even if you have not yet reached your goal.

Possible Causes

a. *Technical glitches that resulted in your injecting a nontargeted ES cell clone.* It is common that the lids of 96-well plates get switched, the labels rub off, the wrong clone is recovered, or the Southern blot is misinterpreted. Go back over all of your data relating to genotyping the putative targeted ES cell clones and regenotype the clone you injected. If the clone turns out not to be targeted, determine all the possible ways that you might have mixed up a genuinely targeted cell clone with a nontargeted one. If you have been taking careful notes and proceeding systematically, you might still be able to find the genuinely targeted clone in the freezer.

If it seems that the clone with which you are working was indeed correctly targeted when it was isolated, there is the remote possibility that a rare wild-type contaminant cell overgrew the targeted cells or that a gene conversion event and subsequent selection for the revertant cells took place during culture.

> **HELPFUL HINT**
>
> *Contamination of a targeted ES cell clone with wild-type cells can be assessed by examining the Southern blot used to identify the homologous recombinants. Both the wild-type and targeted bands should be of comparable intensity. If the intensity of the wild-type band is stronger than the targeted band, then you may have contamination of wild-type cells. One way to prevent overgrowth of contaminating or revertant wild-type cells is to continue ES cell culture in selective medium until shortly before making chimeras; however, this may adversely affect the efficiency of chimera formation.*

You could take a tail sample from the chimeras to do a Southern to determine if the cells used to make the chimera were targeted. However, the most effective way to prevent such problems is to regenotype cell clones as you inject or aggregate them, a smart practice that could be made part of the regular chimera-making routine. Simply collect some ES cells at the end of the experiment and run them through your genotyping protocol. If you have been injecting wild-type cells, you can abort the experiment and go back to *gene-targeting Step 3* or *4* to either isolate new ES cell clones or work with other clones that you know are targeted.

b. *Intended but possibly unpredicted consequences of the genetic alteration, or in other words, a phenotype.* The mutant-bearing spermatids (postmeiotic) or later gametes from the chimera, or possibly the heterozygous embryos resulting from the chimera test-mating, might not be surviving due to a lack of the gene product in the gametes or haploinsufficiency in the case of embryos. For further analysis of this kind of phenotype in the haploid gamete or heterozygous embryos, you will be dependent on having a steady supply of good, germ-line transmitting chimeras. Therefore, after you have established that this is indeed the situation (see Chapter 7), you may need to go back to *gene-targeting Step 4* to make more chimeras, repeatedly (see Box 3.9).

c. *Yet another phenotype.* With 129 ↔ C57BL/6 chimeras, it is common to assess germ-line transmission at 7 days postnatally when agouti hairs become apparent. If you have a dominant perinatal lethal mutation, you might miss the heterozygotes if you rely on coat color as the first indication. Check for perinatal losses and if they are present, genotype the dead pups to see if these are the missing heterozygotes. If this is the case, you have a phenotype and a dominant effect. Go to Chapter 7.

> **Box 3.9** Circumventing the Problem of a Haploid Effect on Sperm
>
> Having a mutant allele that detrimentally affects haploid gametes is a difficult but not impossible situation for analysis, since you may need to have a steady supply of chimeras to produce the defective gametes for study. There may be some hope of circumventing the problem, however, if the gene affects only spermatogenesis. The mutant allele could then be transmitted through the female germ line. In this situation, the choices are to (1) screen all of your targeted clones for the Y chromosome in the hope that one of them will have lost it and could therefore produce female chimeras with a good chance of transmitting the targeted allele; (2) test-breed all of your female chimeras anyway, on the chance that one of them might produce oocytes from the XY ES cells (transmission of XY ES cells from females has been often reported but it is still unclear whether XY oocytes form or whether the Y chromosome is lost somewhere along the line, leaving XO oocytes); or (3) start over at *gene-targeting Step 2*, this time using an XX ES cell line. If transmission is successful through the female germ line and heterozygous mice can be obtained, the effects of the mutation on spermatogenesis in heterozygous male pups can be studied by backcrossing heterozygous females to wild-type males.

3.D Are the Heterozygous Offspring of the Chimeras Normal, Fertile, and of Both Sexes (Box 3.10)?

YES	Congratulations! You are on your way and can concentrate on analyzing a homozygous mutant phenotype in the next generation. Go on to Chapter 4 and eventually Chapter 5 ▶
NO	Congratulations! You have a phenotype ▼

Possible Causes

a. *Dominant mutation.* If the heterozygous mice make it to term, the heterozygous mutation is clearly compatible with embryonic development. However, a dominant phenotype could manifest as abnormalities at birth or at any stage thereafter. Watch for postnatal heterozygote-associated abnormalities and analyze the mice as described in Chapter 6 for homozygous mutants (also see Chapter 7).

b. *Dominant effect on gametogenesis.* If the male and female heterozygotes are infertile, but otherwise morphologically normal, this could be an indication that your mutant has a dominant effect on gametogenesis or some aspect of reproductive behavior. However, because you get the mutation through the germ line of either a male or female chimera in the first place, it has to be something that can be rescued by coexistence with wild-type cells in a chimera (male or female, depending on your breeding scheme), at least in that sex. Also, there are many reasons, other than a possible effect of your mutation, that a mouse might be infertile and it is best to breed a reasonable number before coming to this conclusion. If infertility is always associated with the mutation in one sex or the other, or both, this is the phenotype, and you can go to Chapter 7 (and Chapter 8) for help.

> **Box 3.10** Identifying Male and Female Mice
>
> How do you determine if your mouse is male (has testes) or female (has ovaries)? We use this definition of males and females because some mutants have testes but are genetically XX or have ovaries but are genetically XY. Typically, you assess the phenotype by examining the anatomy of the external genitalia of the mouse. The genetic sex of a mouse can be determined by DNA genotyping for Y-chromosome-linked genes such as *Sry* or *Zfy* (see Box 3.7).
>
> Male and female mice can be distinguished at birth. The distance between the genital papilla and the anus is longer in males than in females (Figure 3.5A–D). Males also have pigmentation in this region unless you are working with an albino strain (Figure 3.5C). The simplest way to assess the sex of newborn mice is to sort the pups in the litter, comparing the genital region of each mouse to others in the litter. Usually, you will be able to sort the litter into two groups, males and females.
>
> Once the mice develop their fur, other external anatomical features can be used to distinguish males and females. As the pups age, the difference in the distance between the genital papilla and anus becomes more obvious (Figure 3.5E–J). Mammary gland/nipple formation on the ventral side of the body can be used to identify females since male mice do not have nipples because of the effects of androgens during development. Look at the underside of the mouse for two parallel rows of five spots each in females where the nipples have formed (Figure 3.5E,F). Nonagouti black female mice, such as C57BL/6, have light-colored fur surrounding the nipples next to the black fur, making it very easy to see. This is also true for the commonly used agouti pigmentation alleles. For albino females, the nipples will be pink next to the white fur. Once the mice are sexually mature (~6 weeks old), it is very easy to distinguish males from females (Figure 3.5G–J). If you have any doubts, simply compare the mouse in question with a known male or female (see Box 3.7).

c. *Sex-limited dominant effect.* If heterozygotes of only one or the other sex have physical, growth, or survival problems postnatally, consult a good reproductive biologist or endocrinologist to determine how the effect is limited by the sex of the heterozygote. If fertility is the problem and heterozygotes of only one sex are fertile, the mutation may be specific to spermatogenesis or oogenesis, or to some aspect of reproductive behavior, in which case you can maintain the mutant by breeding heterozygotes of the opposite sex. We doubt that the majority of investigators will ever reach this point in the diagnosis, because these effects should be rare, but for those who do, consider collaborating with investigators interested in gametogenesis or reproductive behavior.

d. *Dominant sex reversal.* If the heterozygotes are all female, it is possible that your mutation causes dominant sex reversal, resulting in XY females. The heterozygous XY female offspring will probably be sterile but the heterozygous XX females will probably be fertile. Thus, minimally, you should be able to maintain the mutation through the female germ line. Making the mutation homozygous will be a challenge (see Chapter 4 for breeding tips and Chapter 7 for further analysis).

> **HELPFUL HINT**
>
> *Genotype the first heterozygous offspring from chimeras by Southern blot analysis with an external probe just in case there was a mix-up between isolation of the targeted ES cell clones and making the chimeras. Once you have confirmed the correctly targeted genotype of the heterozygotes, use a quicker PCR strategy for genotyping mice in subsequent generations.*

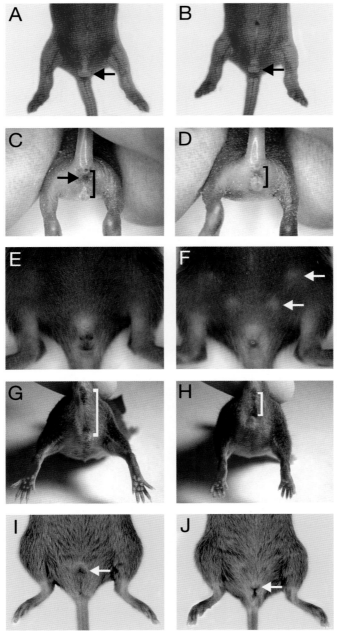

Figure 3.5. External morphology of male and female mice. Ventral view of newborn male (*A*) and female (*B*) mice. Photographs show slight differences in the genital papillae of male and female neonates (*arrows*), and the papilla of males is larger. Rear view of ~2 days postpartum male (*C*) and female (*D*) mice. The distance between the genital papilla and the anus (*brackets*) is longer in males than in females. In pigmented strains, the scrotal area of the male neonates has pigment (*arrow*). Ventral view of ~10-day-old C57BL/6 male (*E*) and female (*F*) mice. At this age, the nipples of the female can be seen because of the lighter surrounding hair (*arrows*). Rear view of male (*G*) and female (*H*) adults. The genital–anal distance (*brackets*) is much longer in males compared with females. Ventral view of male (*I*) and female (*J*) adults showing the gross differences in their external genitalia. Arrows point to the genital papilla.

CHAPTER 4

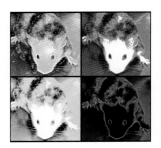

Maintaining Your Mutant

ALTHOUGH THE POSSIBILITY OF PRODUCING A DOMINANT MUTATION has been raised repeatedly in Chapter 3, do not get the wrong impression: By far, the majority of targeted mutations, or spontaneous ones for that matter, turns out to be recessive. Thus, it is highly likely that successful production of a chimera that transmits the mutant allele through the germ line will result in viable, fertile, heterozygous male and female mice. The next step in the gene-targeting experiment is to name your mutation (see Box 4.1). Then you will mate heterozygotes to produce homozygous mutant offspring, which will be expected to have a phenotype. But first, let us digress for a moment to consider the best strategies for this all-important mating and for maintaining this hard-won mutant. If you are a veteran mouse breeder with a background in genetics and experience with mutant mice, you can skip this chapter. If you have a dominant phenotype and were sent here from Chapter 7, look through this chapter for hints on breeding in general and specific tips for the special situation of breeding a dominant infertility mutation (Section 4.D.3).

HELPFUL HINT

If at any point you either lose interest or can no longer afford to keep your mutant, but you think it might be useful to someone else, consider preserving it as frozen embryos (MM3)* or frozen sperm (MM3) (so that in vitro fertilization can be done at a later date), submitting it to a mutant repository (see Appendix 1), or at the very least, giving it to an interested colleague.

---

*MM3 refers to *Manipulating the mouse embryo: A laboratory manual*, 3rd edition (Nagy et al. [2003]. Cold Spring Harbor Laboratory Press, Cold Spring Harbor, New York).

> **Box 4.1** How To Name a Mutant Allele
>
> The descriptive, sometimes fanciful names given to mouse mutants in the past, such as *varitint waddler, kinky waltzer,* or *screw tail,* are gradually giving way to names based less on the phenotype of the mutant and more on primary features of the gene product of the wild-type allele. Many genes have changed names as the genes have been cloned and their protein products discovered. A familiar example is the *albino* mutation named to describe an all-white mouse; the locus is now known as *tyrosinase* (*Tyr*) after the enzyme that the gene encodes, and the descriptive mutant allele names such as *albino, himalayan,* and *chinchilla* are now identified by superscript symbols (*Tyr^c*, *Tyr^{c-h}*, *Tyr^{c-ch}*).
>
> When naming a new mutation in a known gene made by gene targeting, you will be limited to using the accepted gene name and symbol, and your main concern will be in adding the superscript "tm" for targeted mutation and providing a unique identifier for the allele (Figure 4.1). The rules for deciding new gene names are made by a nomenclature committee (http://www.informatics.jax.org/mgihome/nomen/index.shtml) and there is not much room for creativity. However, you may personalize the allele name by adding your own unique laboratory or institution code. If you do not have a laboratory or institution code, you can get one from the Institute for Laboratory Animal Research at http://dels.nas.edu/ilar/codes.asp?id=codes/. Note that many journals require that new allele names be approved by the nomenclature committee before publication.
>
> If the mutant allele you are analyzing was created as a result of random insertion of a transgene, i.e., insertional mutagenesis, the mutation will be named as a mutant allele of the gene that was disrupted, and the superscript will contain the symbol "Tg" for transgene. The name of the gene that was inserted is added in parentheses but can be omitted if the mutation is uniquely identified without it. Finally, the gene name can be attached to a strain designation to indicate the genetic backgound on which it is maintained (Figure 4.1).

4.A Animal Facilities

By far, the preferred animal facility is a specific pathogen-free (SPF) barrier facility. The specific pathogens that are excluded from such facilities include parasites, certain bacteria, mycoplasma, and viruses. SPF facilities, rapidly becoming the accepted standard for research institutions, are facilities that you should promote if they are not already in place at the institution in which you intend to maintain your mutant mice.

An SPF facility uses procedures and equipment for animal housing, handling, and husbandry that maintain animals in a SPF environment. They also rigorously monitor incoming mice, tumors, cell lines, equipment, and personnel to prevent the introduction of pathogens into the facility. These procedures may require animal quarantine or rederivation of mice by Caesarean section before importation, often at considerable expense and sometimes inconvenience for the investigator. However, these procedures are well worth the trouble in the long run, and investigators should insist on strict adherence to the rules.

The advantages of working with SPF mice should be self-evident, but they bear repeating. Animals infected with pathogens may have increased mortality and impaired breeding performance, necessitating a much larger number of animals for both maintenance and experiments than if they were healthy. The effects of endemic pathogens may be transient or subclinical, but will have effects on the host immune system that could invalidate research data. Variations in the population of mice due to the natural course of infections could lead to variations that confound the phenotypic analysis or that could be mistaken for a phenotype. Mice with poor health status also present a roadblock for collaborations, since they will have to be rederived or quarantined before import into any SPF facility. All in all, healthy mice make better research subjects and are the basis for accurate phenotypic analysis of genetic mutations.

4.B Genetic Background

Among strains of mice, many differences are readily apparent: color, behavior, size, and shape, to name a few variable characteristics. The basis for these strain differences in phenotype is allelic variation in genes, the so-called background genotype. It is against this background that major gene effects—spontaneous or induced single gene mutations—are analyzed. Because no gene functions in isolation and a phenotype is the product of the gene's action and the action of all other genes affecting the system, it stands to reason that single gene effects, however large, are subject to variation in phenotype, depending on the background genotype on which they are examined.

Most likely, your choice of mouse strain for test-breeding of chimeras is based on some combination of breeding performance, convenience, or useful markers (see Box 3.5, p. 38). Therefore, your heterozygous mice will most likely be either F_1 (first filial generation) between the embryonic stem (ES) cell strain (usually 129) and another inbred strain, such as C57BL/6, or they will be half 129 and half something else, such as outbred, F_1, or some other mixed genetic background. All of these mixed-genotype possibilities are advantageous for rapidly expanding the colony of mice carrying the mutation, because all of these combinations will have the advantage of hybrid vigor. Even if the homozygous mutation is lethal, you should certainly not lose the mutation if you are careful. The only drawback of this arrangement is that the first homozygous mutant mice will inevitably have a mixed genetic background, and there could be genetic variability affecting the mutant phenotype (see Box 4.2). In other words, in

---

Box 4.2 Genetic Background Effects: Modifiers of Phenotype

For the most part, the phenotypic effects of a gene modification will be major compared to normal developmental variation and they will produce a readily detectable departure from normal. However, the phenotype may have variable penetrance (appearing in only some individuals) or variable expressivity, that is, it may vary considerably from one individual to the next. The source of this variability could be either developmental or genetic. Such genetic effects on the phenotype from genes other than the mutated locus are known as genetic background effects and the genes that cause them are genetic modifiers. They can be revealed by the observation that the penetrance or expressivity of the phenotype is different when observed on different genetic backgrounds, i.e., on different strains, and they may be the result of one or many genes in which the two strains have different alleles. Typically, genetic background effects are minor and there is usually not much interest in chasing down exactly what other gene(s) is causing the variation; it could be a single gene with a minor effect or a combination of many genes each with a small effect. It is enough to know that the phenotype is subject to background variation. Occasionally, however, a radically different phenotype may result when a mutation is placed on a different genetic background, making it much more enticing to map and identify the specific gene modifier(s) as an important contributor to the phenotype. It could be that an allele at a single locus acts as a major genetic modifier of the phenotype and is therefore of significant biological interest to the study of your favorite gene. Thus, whether or not there seems to be variation in the phenotype of your mutant, you could have a surprise when it is examined on a different genetic background.

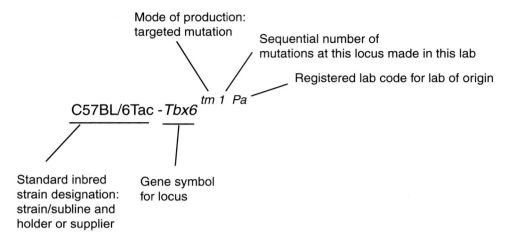

Figure 4.1. Naming a new allele made by targeted mutagenesis. A superscript is added to the gene symbol of the mutated gene to provide a unique identifier: tm indicates that the allele was made by targeted mutagenesis, and a number and laboratory code identify the mutation and indicate who made it. The allele name, in italics, is preceded by the name of the strain of mouse on which the mutation is maintained. This example, C57BL/6Tac-*Tbx6^{tm1Pa}*, designates the first mutant allele of the *Tbx6* locus made in the Papaioannou laboratory using targeted mutagenesis, maintained as a congenic strain with the C57BL/6 inbred strain from Taconic Farms.

gaining the advantage of hybrid vigor, the advantage of a uniform genetic background will be lost in this first analysis. Of course, this is not the case for heterozygous effects that will be analyzed initially on an F_1 background in the first generation (the offspring of the chimera and an inbred test-breeder), since F_1 mice between two inbred strains are all genetically identical.

If you begin the analysis of your mutant on the mixed background (and it is difficult to resist the temptation to begin immediately), start to transfer your allele to one or more inbred strains as outlined in the following paragraphs. This will allow you to look at the mutant effect on different genetic backgrounds, potentially uncovering genetic modifiers of the targeted gene and possibly reducing phenotypic variability. The choice of strain may be dictated by the expected phenotype; for example, if you are working with a suspected tumor suppressor gene, you might choose an inbred strain with a high susceptibility to tumors. Or the strain might be chosen for convenience, such as availability of genetic sequence information, or there may be no rational reason for the choice because there is not necessarily any sure means of predicting which will be a useful strain/mutant combination.

If the ES cells you used for gene targeting were from an inbred strain (such as 129), then breeding a germ-line-transmitting chimera to mice of that same strain is the quickest way to place the mutant allele onto an inbred background. In the first generation, you will have a co-isogenic line (i.e., genetically identical to the parent inbred strain except for the altered allele). If you used a 129-ES cell line, however, be aware that there are many different 129 substrains (see Box 4.3). If mice of the specific substrain from which the ES cells were derived are not available for breeding, then any cross with a different 129 substrain will not produce a true inbred strain, but rather a mixed substrain. However, for all but the genetic purists (and we admit to being so inclined) the resulting mice will approximate an inbred, co-isogenic strain.

Another good choice of inbred strain on which to transfer your mutation is C57BL/6, not because this strain is particularly easy to work with, but simply because these mice are widely used

Box 4.3 The 129 Substrain Morass Made Easy

ES cells from the 129 strain of mice are most often used for gene-targeting experiments because of the ease of obtaining the ES cells from embryos of this strain and because of their propensity to colonize the germ line in chimeric hosts. For efficient gene targeting, it has long been recognized that using isogenic DNA in the targeting construct is of prime importance. However, in terms of genetic purity, the 129 strain could hardly be a worse choice. The strain originated in 1928 from crosses of several fancy coat color mice. From that cross, a variety of 129 substrains were developed either intentionally or simply by random genetic drift following separation of breeding colonies, and various contaminating crosses were made either intentionally or inadvertently. The result is a range of 129 substrains differing or segregating for a variety of genes, with whole chromosomal segments derived from other strains. This becomes important to the gene targeter when using a genomic library from one 129 substrain (e.g., the Stratagene 129 library, which is derived from the 129/SvJ substrain) for producing a targeting construct, and using ES cells from a different substrain (e.g., R1 ES cells, which are derived from a cross between 129/Sv and 129/SvJ substrains) for gene targeting (see Table 4.1). Low targeting efficiency may be a problem for specific loci that have a different origin in the genomic library and the ES cells, i.e., loci that are nonisogenic. Fortunately, the variant chromosomal stretches have largely been mapped, and it is possible to identify those loci that are likely to be more difficult to target (see Appendix 1). It is worth finding out just which substrains you are dealing with, for both the ES cells and the genomic library used for obtaining clones, to determine whether the gene you are targeting falls into one of the contaminated regions.

Table 4.1. Derivation of some common 129-ES cell lines with the old and new 129 substrain nomenclature (http://www.informatics.jax.org/mgihome/nomen/strain_129.shtml) and the relevant coat color alleles that these ES cell lines carry

ES cell line	Substrain of embryo of origin		Relevant coat color alleles in ES cells
	Old nomenclature	New nomenclature	
AB1	129/SvEvBrd	129S5	A^w
CCE	129/SvEv	129S6	A^w
E14TG2a[a]	129/OlaHsd	129P2	A^w; Tyr^{c-ch}; p
J1	129/SvJae	129S4	A^w
R1	129/Sv x 129/J	129S1 x 129X1	A^w; Tyr^{c-ch}; Tyr^+; p; p^+

[a]The ES cell-derived component of chimeras made with this cell line will be very pale colored; all the others will be agouti.

in research, are available in large numbers from commercial sources, and many mutations are maintained in this strain. In addition, their genome has been sequenced, assembled, and notated.

If you used the common scheme of injecting targeted 129-ES cells into C57BL/6 blastocysts and test-bred chimeras to C57BL/6 females, then in the first generation you are already

HELPFUL HINT

Note that C57BL/6 females have a high incidence of septate vaginas, sometimes approaching 10%. This is apparent as bilateral vaginal openings separated by a membranous septum rather than a single vaginal opening. In our experience, these females are not good breeders. It is probably best to cull them at weaning.

halfway there, genetically speaking, with (C57BL/6 × 129)F_1 mice. Getting the rest of the way to having a congenic strain (i.e., identical except for a differential chromosomal region surrounding the mutant allele) is a simple matter of backcrossing heterozygous animals to C57BL/6 mice with forced heterozygosity at your mutant locus for ten generations (the original cross is the first). This is more time-consuming than getting the mutation onto 129, although it involves very little work (see Box 4.4), so you might as well get started right away. By the time the first manuscript reporting the phenotype of your mutant on a mixed genetic background is ready, you might be in a position to look at a few mutants on a different genetic background to determine if there are major variations in the phenotype.

> **Box 4.4** Backcross Breeding Scheme to Make a Congenic Strain
>
> To put a mutant allele onto an inbred background, cross a mouse heterozygous for the mutation to a mouse of the chosen inbred strain. This is called an outcross. If the mutation is originally on an inbred background (e.g., coming from a chimera made with 129-ES cells), then the offspring of the outcross will be genetically uniform F_1 mice, with half of their genome coming from the 129 strain and half coming from the other inbred strain. By the way, the genetic background of this F_1 generation can always be recreated after a congenic strain is obtained by outcrossing with 129 mice. If the mutation is originally on a mixed genetic background, as might be the case for a spontaneous mutation, the outcross creates a unique background that is 50% from the mixed background and 50% from the inbred background. In the next cross, and in all those that follow, mice selected for heterozygosity at the mutant locus are backcrossed to mice of the inbred strain. At each generation (called generation $N_2...N_n$), the genetic diversity is decreased and the mice contain more and more of the inbred genome: at N_{10}, this reaches 99.8% and the mutant strain is considered congenic with the inbred strain (Figure 4.2). Theoretically, the mice are genetically identical except for a differential chromosomal region of about 20 cM (~40,000 kb) containing the mutant gene. Because heterozygotes were selected at each generation, the chromosomal region around the mutation was selected, and it will only become smaller by random crossover during meiosis. It would take another 40 generations of backcrossing to reduce the size of the differential chromosomal segment to less than 5 cM by random crossover. If you are motivated, you can speed up this process by screening N_{10} heterozygous backcross mice for crossovers close to the targeted allele using DNA markers. Use these selected heterozygotes for the next backcross and then repeat the process on the other side of the locus in the next generation. However, for most practical purposes, accepting 99.8% genetic uniformity is a fairly good alternative. In fact, you might want to monitor the phenotype during earlier generations of backcrossing because an effect might be seen quite early on; by N_4, the mice are already 87.5% identical to the inbred strain.
>
> Congenic mouse lines can be generated more quickly using the "speed congenic" method. At each backcross generation, mice are genotyped for genome-wide polymorphic markers. The mice that have the highest percentage of markers from the chosen inbred background are then selected for use in the next backcross generation. This selective breeding reduces the number of generations needed to generate a congenic line but requires more work than simply breeding mice.
>
> When backcrossing, it makes sense to use a male of the inbred strain, because you can use the same mouse for several generations if he is a good breeder. Simply mate him with his heterozygous daughters or granddaughters. If instead you choose to use inbred females, just be sure to use an inbred male for at least one generation to get the Y chromosome of the inbred strain into your line. You may notice a decline in fertility as you progress through the generations. That is because you started with an F_1 (or a mixed genetic background) that had hybrid vigor and you are moving toward a genetically uniform inbred background. However, the fertility should approach that of the inbred strain and not get any worse, unless, of course, the mutation has a heterozygous phenotype affecting fertility.

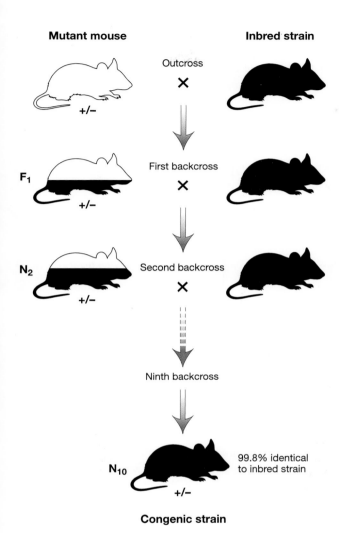

Figure 4.2. Backcross scheme to make a congenic strain. By crossing a heterozygous mutant to an inbred strain and selecting heterozygous offspring from the F_1 generation to cross back to the inbred strain for the next generation (N_2), a congenic strain that is 99.8% identical to the inbred strain but carries the mutation will be achieved after the ninth backcross.

4.C Maintaining a Mutant Colony: The Basics

It is useful to think about the breeding of your mutant mouse colony as having two components: (1) maintenance of a production colony to perpetuate the mutation and (2) production of mutant mice or embryos for phenotypic analysis.

4.C.1 Maintaining a production colony to perpetuate the mutation

Your object is to keep the mutation that you have produced in a live animal as a permanent resource for yourself and the scientific community. Unless the mutation is severely detrimental to the health or reproductive capacity of the heterozygous mice (see Section 4.D.3), this is usually quite straightforward using basic husbandry techniques as approved by your local

Institutional Animal Care and Use Committee (IACUC) for housing, genotyping, and marking individual mice (MM3). We routinely take the tip of the tail for genotyping either at 7–10 days postnatally or at the time of weaning. The advantage of the earlier date, especially if space is limited, is that genotpes will be known by the time of weaning and unneeded mice can be culled. Waiting until weaning requires one less visit to the mouse room, but all mice will have to be weaned and stored until genotypes are known. If the mutation is dominant or semidominant, the genotyping task may be easier, provided that the phenotype is fully penetrant, i.e., all of the heterozygotes show the effect. If you can distinguish every heterozygote by gross morphology or behavior, you can dispense with the genotyping step and simply breed the animals with a mutant phenotype to phenotypically (and thus genotypically) wild-type mice.

HELPFUL HINT

Sometimes, disasters do strike animal facilities. As insurance, send some mutants to a collaborator in a different institution and/or submit them to a mutant repository (see Appendix 1). Alternatively, freeze some sperm or embryos (MM3) for safekeeping and archive them in more than one location.

When maintaining a colony on a mixed or outbred genetic background, beware of inbreeding depression. If you start with a small number of heterozygous mice and intercross them in a small closed colony with forced heterozygosity (i.e., select heterozygotes for breeding), you will inbreed at a fairly rapid pace. What was originally a vigorous breeding colony may quickly lose reproductive potential and threaten to die out. Inbreeding depression can actually be more of a problem when you start with an outbred population than when you work only with inbred mice. This is because inbreeding depression is the result of uncovering detrimental recessive alleles that segregate in a population. In the case of an outbred stock, the effects will be particularly severe between the second and eighth generations of inbreeding as the deleterious alleles become homozygous. In the case of starting with a cross between inbred strains, all of the really deleterious alleles will already have been eliminated from the population during the initial derivation of the inbred strains.

This inbreeding depression has nothing (necessarily) to do with your mutation, but effects of the mutation could exacerbate the problem. The solution is quite simple: Bring in new blood. This can be accomplished simply by (1) producing additional vigorous hybrid heterozygotes by continuing to breed the germ-line chimeras or (2) outcrossing the heterozygotes to an inbred strain or even an outbred stock of mice.

Remember that for expansion of a production colony, crossing a heterozygote with a wild-type mouse—useful if the heterozygotes are in demand for experiments—produces exactly the same proportion (50%) of heterozygous mice as crossing two heterozygotes together. Furthermore, with a heterozygous × wild-type cross, the genotyping is simplified because you need only distinguish between two genotypes. In the event that the mutation has no effect on fertility of the homozygous mutants, these mice can be mated with wild type, and 100% of the offspring will be heterozygotes (with no genotyping necessary). However, do not be tempted to keep the production colony by breeding only homozygous mutants, since you always want to keep the wild-type allele in the colony for producing wild-type littermates with the same or similar genetic background for experimental controls.

In addition, maintaining a mutation on an inbred background should not create problems, although you will almost certainly need to maintain a larger number of breeding pairs than

> **HELPFUL HINT**
>
> *As you expand your colony of mice carrying the mutation, you will likely be breeding heterozygous males with wild-type females and heterozygous females with wild-type males. If you notice a mutant phenotype in the resulting heterozygotes, but only from one of these types of crosses, then you have evidence for an autosomally imprinted mutant phenotype (see Section 7.E).*

those of mixed background to produce a similar number of offspring. If you produced an "instant" inbred (co-isogenic) strain by, for example, breeding 129 inbred mice to a 129-ES cell germ-line chimera, the reproductive characteristics of the colony should be the same as those of the 129 inbred strain, and you do not have to worry about any further inbreeding depression since deleterious alleles will have been eliminated from the strain during its original derivation.

> **HELPFUL HINT**
>
> *Do not allow your stocks to get stale! Maintaining a colony usually involves keeping a certain number of "stock" cages of genotyped mice for future use. Keep a watchful eye on these cages and replace stock mice, as they age, with weanlings.*

The mutation is maintained by forced heterozygosity and the strain is called a segregating inbred strain. If you are backcrossing your mutation onto a different inbred strain (see Box 4.4), however, you will have hybrid vigor in the early generations that gradually disappears in the later generations as the congenic strain approaches genetic uniformity with the inbred strain. Be careful of this loss of productivity and set up more mating pairs as the reproductive performance declines. Your goal in the production colony is to produce a constant supply of heterozygous and wild-type animals in sufficient number to replenish breeders as they age or their fertility declines, and to supply plenty of heterozygous animals for the mutant analysis. In the case of a homozygous lethal mutation, a heterozygous female will have to be sacrificed for each litter studied, so the number required can be considerable.

4.C.2 Production of mutant mice or embryos for analysis

From your actively breeding mutant-maintenance colony, production of mutants for analysis is straightforward. If male and female heterozygotes are fertile, mate them together: One quarter of the offspring will be homozygous mutants; one-half will be heterozygotes, which might have a phenotype if the mutation is dominant or semidominant; and the remaining quarter will be wild type and can serve as controls. Set up some heterozygous males in individual cages to serve as studs in sufficient number to supply your experimental needs. Once their genotype and fertility have been confirmed by the production of homozygous mutant offspring, these males become priceless. Leave them in their home cage, mate them by sequentially introducing heterozygous females, and remove the females when they have mated (i.e., when a vaginal plug has been detected [see Figure 3.2]). This scheme will produce a steady supply of pregnant females for the analysis of offspring or embryos. As a standard safeguard to ensure that the female is really a heterozygote, take a tissue sample to regenotype each of the mothers you use to generate embryos or offspring for experiments. That way, your Mendelian proportions will not be skewed by inclusion of litters from incorrectly genotyped mothers. This is especially important early on when you are still determining when your mutant phenotype shows up and whether it is variable in presentation.

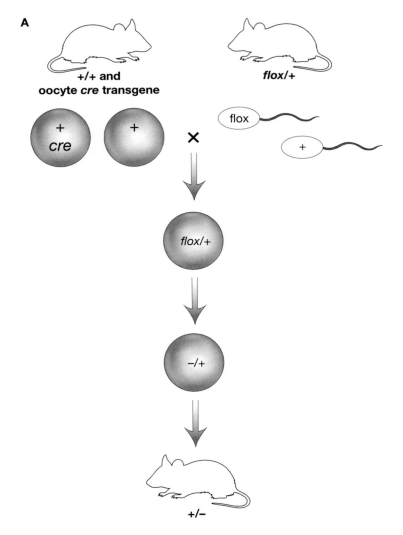

Figure 4.3A. (*See facing page for legend.*)

4.D Maintaining a Mutant Colony: Beyond the Basics

4.D.1 Deletion of selection cassettes in vivo

If you generated a targeted mutation in ES cells, you used a drug-resistance expression cassette in the gene-targeting strategy to isolate DNA transformants and identify homologous recombinants. These cassettes usually have their own promoters, which can potentially influence the expression of neighboring genes and/or the expression of your targeted locus. Therefore, in most cases, it is desirable to remove the cassette from the targeted locus. This is relatively simple using the Cre/*loxP* or Flp/*FRT* recombinase systems (Section 2.A.5). Floxed or flrted cassettes can be removed either in vitro in the targeted ES cells before making chimeras (Box 2.3, p. 17) or in vivo by crosses with mice that express the appropriate recombinase (Figure 4.3). Floxed

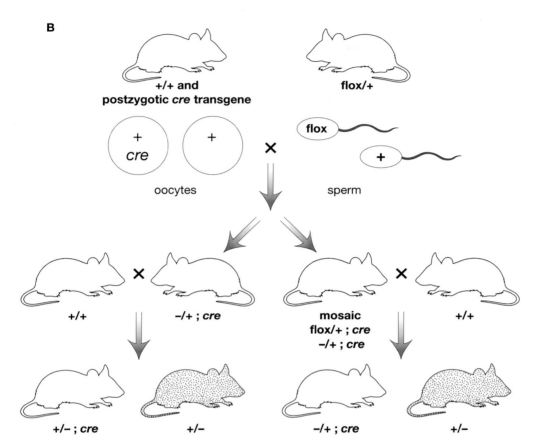

Figure 4.3. Deletion of selection cassettes in vivo. (*A*) Oocyte-Cre deleter strategy. Transgenic mice express Cre during oogenesis, before the final meiotic divisions. Two types of oocytes will be generated: those that carry the *cre* transgene and those lacking the *cre* transgene. However, all of the oocytes will have Cre protein (*shading*). When sperm from a male carrying a floxed drug-selection cassette fertilize these oocytes, the floxed cassette will be deleted in the zygote even in the absence of the *cre* transgene. Mice heterozygous for the cassette-deleted allele and without the transgene are then selected to perpetuate the mutation. This strategy is attractive because you delete the drug-selection cassette (or any floxed sequence) and get rid of the *cre* transgene in one step. (*B*) Postzygotic Cre deleter strategy. Transgenic mice express Cre in blastomeres, the inner cell mass (ICM), or the epiblast before germ cell allocation. Two types of oocytes will be generated from these mice: those that carry the *cre* transgene and those lacking the *cre* transgene. In contrast to the oocyte-Cre deleter strategy, oocytes without the *cre* transgene will not have Cre protein. When sperm from a male carrying a floxed drug-selection cassette fertilize these oocytes, the floxed cassette will be deleted postzygotically only if the *cre* transgene is also inherited. Sometimes this results in a somatic and germ cell mosaic (i.e., some cells retain the drug-selection cassette and some delete the floxed sequence). Mice with the floxed sequence deleted can be recovered by an additional cross to wild-type mice. This also serves to segregate the *cre* transgene away from the recombined allele.

or flrted selection cassettes can be removed in vivo by crosses with transgenic mice that express Cre or Flp, respectively, in germ cells, blastomeres, the ICM, or the epiblast before primordial germ cell formation. These types of recombinase mice are sometimes called deleter mice. The first of two situations to consider is that mice heterozygous for a floxed (or flrted) selection cassette are crossed with mice heterozygous for a *cre* (or *flp*) transgene that is expressed in developing oocytes before meiotic divisions (Figure 4.3A). The deleter female will ovulate two types of oocytes: those that carry the recombinase transgene and those that do not. However, all of the oocytes will have recombinase protein because of the premeiotic expression of the transgene. Thus, when sperm carrying the floxed (or flrted) cassette fertilize the recombinase-carrying oocytes, all of the resulting zygotes (recombinase transgene, positive or negative) will have the floxed (or flrted) cassette removed. This strategy can also be used for crosses between male ES cell chimeras transmitting a floxed (or flrted) selection cassette and deleter females for removal of the selection cassette on germ-line transmission. Similarly, some ES cells have been engineered to express Cre in male germ cells so that floxed cassettes can be removed during germ-line transmission from ES cell chimeras. The only drawback is that the progeny of the chimera will carry the *cre* transgene, which will have to be segregated from the targeted allele by another cross.

In the second situation, mice heterozygous for a floxed (or flrted) selection cassette are crossed with mice heterozygous for a *cre* (or *flp*) transgene that is expressed in blastomeres, the ICM, or the epiblast before primordial germ cell formation (Figure 4.3B). Because recombinase expression occurs after fertilization, deletion of the floxed (or flrted) selection cassette will occur only if the mice carry both the recombinase transgene and the floxed (or flrted) selection cassette. Therefore, to generate mice carrying the targeted mutation without the selection cassette, you will need to breed the double heterozygotes with wild-type mice to segregate the deleted allele and the recombinase transgene.

If you want to remove a floxed selection cassette and simultaneously maintain your targeted allele on a 129 inbred genetic background, then you must have a Cre-expressing mouse of the same 129 inbred genetic background, which can be hard to find. 129 inbred Cre deleter mice have been reported, but with so many different 129 substrains (see Box 4.3) that generating a true co-isogenic strain will be limited by *cre*-transgenic mouse strain availability. A solution to this problem is to generate inbred zygotes that contain your targeted allele with the floxed selection cassette and then to microinject pronuclei with a Cre expression plasmid—a circular form for transient expression without integration. The resulting progeny may have the selection cassette removed in all tissues or they may be mosaic. If they are mosaic, you can backcross to the 129 strain to segregate the alleles. The same strategy can be used for a flrted selection cassette.

If you have a triple *loxP* allele and you would like to remove the selection cassette, you can cross the mice with deleter mice that express weak or variable levels of Cre to generate double heterozygotes. These double heterozygotes are likely to be mosaic for the four possible genotypes generated by different recombination events: no sequences deleted, all floxed sequences deleted, deletion of floxed exons but retention of the selection cassette, and deletion of the floxed selection cassette only (the desired event). Screen the mice by Southern blot analysis to identify mice (preferably males because breeding is more efficient) that have a reasonably strong band for the conditional allele without the selection cassette. Then breed the males with wild-type females to segregate the genotypes in the resulting progeny. If you have used the *loxP/FRT* system (Section 2.B.4), it will be straightforward to remove the selection cassette because all you have to do to generate your conditional allele is breed the heterozygotes with a Flp deleter strain.

4.D.2 Maintaining a mutation using a balancer chromosome

If you have a recessive lethal mutation, you must genotype heterozygotes at each generation to maintain the line. Balancer chromosomes, which have recently been engineered in the mouse, contain large (cM) inversions that suppress recombination over the region of the inversion. In addition, the inversion is marked with a visible dominant marker (e.g., coat color) and a linked recessive lethal mutation. Thus, mice heterozygous for a balancer chromosome can be distinguished from wild-type mice by visual inspection for the dominant marker, whereas mice homozygous for the balancer chromosome die as embryos. To balance a recessive lethal mutation, breed mice heterozygous for your mutation to mice carrying a balancer chromosome with an inversion that spans the region containing your gene mutation (Figure 4.4). Then select males and females that are heterozygous for your mutation and the balancer chromosome. Breed these mice to generate progeny. Progeny homozygous for your gene mutation will die due to the mutation's effects, as will mice homozygous for the balancer chromosome because of the recessive lethal mutation included on it. Therefore, the only mice born from this cross will be mice heterozygous for your gene mutation and heterozygous for the balancer chromosome, i.e., they will have the same genotype as their parents and it will not be necessary to genotype them. A balancer chromosome does not necessarily have to have a linked recessive lethal mutation if the dominant visual marker can be used to distinguish mice heterozygous or homozygous for the balancer.

Balancer chromosomes can also be used to maintain a recessive infertility gene. As with the lethal mutation, mice homozygous for the balancer chromosome will die and mice heterozygous or homozygous for your mutation and the balancer gene will be distinguishable by the visible marker on the balancer chromosome. No genotyping or test-breeding is necessary. At the moment, only a few balancer chromosomes are available in mice—covering a minority of the genome—but their numbers are increasing rapidly.

4.D.3 Assisted reproduction

It may be that your mutant gene causes dominant infertility or that you simply find yourself in the situation of having the mutant gene in only a few mice that are infertile, even though this might not be part of the mutant phenotype. In either case, to rescue the mutation, a series of increasingly drastic steps can be taken to assist reproduction. Of course, if the infertility is limited to only one sex, breeding heterozygotes of the fertile sex to wild-type mice can perpetuate the mutation. But when you want to examine homozygotes, you may need to rely on assisted reproduction to overcome the infertility of the other sex.

4.D.3.a Dealing with male infertility

If an important mutant male mouse is not breeding, then a series of assisted reproduction methods can be performed to rescue his germ line. First, superovulated (MM3) weanling females can be provided to the male. If a plug is obtained, you can collect the fertilized oocytes and transfer them into pseudopregnant foster mothers to generate progeny (MM3). It is possible that you will recover oocytes but they will not be fertilized even if a plug is present. If this occurs, attempt a second mating and if this does not yield fertilized oocytes, then in vitro fertilization (IVF) or artificial insemination (AI) (MM3) is indicated.

For IVF, isolate sperm from the cauda epididymis of the male mutant. Because each male has two epididymides, one can be removed surgically for IVF without sacrificing the animal, leaving the other intact for a second attempt later. Transfer fertilized oocytes that progress to the two-cell stage after IVF into the oviducts of pseudopregnant foster mothers to generate progeny. At each IVF attempt, sperm can be frozen for subsequent manipulations (MM3).

At this point, if the IVF attempts are unsuccessful, some options are possible. Epididymal or testicular sperm can be used for intracytoplasmic sperm injection (ICSI) into oocytes (MM3). In addition, if the mouse is extremely important, one could consider animal cloning (MM3), although this is impractical as a means of perpetuating mutations. If this is a targeted mutation, it would be much simpler to recreate the mouse the way it was obtained, by breeding the ES cell chimeras. If the infertility is a dominant effect of the mutation, you may never be able to get homozygous mutants by assisted reproduction but the reproductive phenotype can be studied in the heterozygous offspring of the chimeras.

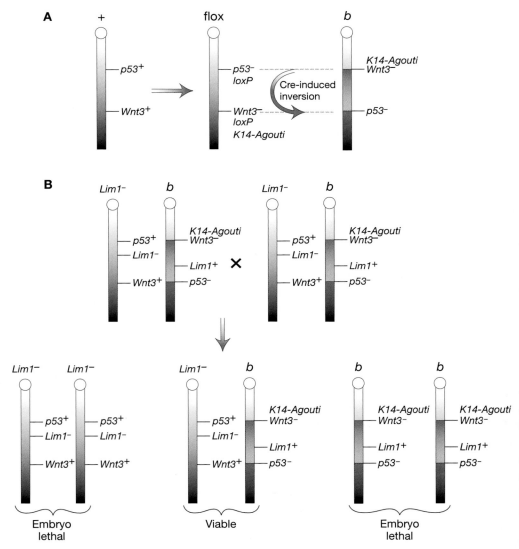

Figure 4.4. (*See facing page for legend.*)

4.D.3.b Dealing with female infertility

If an important mutant female mouse is not breeding, first exchange the male partner for a new, young one (6–8 weeks old) and wait to see if a pregnancy results. If the female does not become pregnant within one month, inject the appropriate hormones for superovulation. If you obtain a plug, collect fertilized oocytes for transfer into recipient females and save the ovaries for ovary transplantation. Ovary transplantation is a standard methodology used for rescuing the germ line of female mice (MM3) that could also be used if no plug is obtained. Ovaries or pieces of ovaries can be transplanted into the bursae of normal histocompatible hosts where they will ovulate functional oocytes and, if fertilized, produce progeny. If your female mutant is on the usual B6/129 mixed genetic background, you can use a B6129F1 female as a histocompatible host. If your mutant is on an outbred genetic background, you will have to consider transferring the ovaries into an immune-compromised female recipient (e.g., a *nude* mouse). Donor ovaries can be isolated from adult mice, prepubertal females, or even newborn females. If the ovary transplant fails, you might want to consider animal cloning (MM3); freeze some tissues (e.g., tail fibroblasts) from the female for this purpose.

Under certain circumstances, other types of artificial reproductive techniques, such as AI, IVF, or ICSI, might be appropriate. As with male infertility, a dominant infertility effect might preclude the production of homozygous mutants, but the reproductive phenotype can be studied in the heterozygous mice.

4.E Special Breeding Techniques for Use in Mutation Analysis

At this point, you are ready to proceed with phenotypic analysis. First, have a look at the following four sections, which outline some special breeding schemes that you might be able to use in the phenotypic analysis of your mutation. Then go on to Chapter 5 or back to Chapter 7 if you came from there.

Figure 4.4. Maintaining a recessive lethal or sterile mutation using a balancer chromosome. (*A*) Diagram of a balancer chromosome generated by chromosome engineering on chromosome 11. Two *loxP* sites in reverse orientation were introduced in *cis* into the *p53* and *Wnt3* loci using gene targeting in ES cells (flox chromosome). Cre expression caused the inversion of this 24-cM region. The targeting strategy also disrupted the *p53* and *Wnt3* genes and introduced a *K14-Agouti* transgene into the *Wnt3* locus. The balancer chromosome (*b*) has (1) a 24-cM inversion that suppresses recombination with the wild-type chromosome (+) within the inverted region, (2) a recessive embryonic lethal mutation, i.e., *Wnt3–*, and (3) a dominant coat color marker, i.e., agouti pigmentation. Mice heterozygous for the balancer chromosome (*b*/+) can be recognized on both agouti and nonagouti genetic backgrounds because they have lighter ears and tails due to *K14* directed expression of *Agouti*. Mice homozygous for the balancer chromosome (*b*/*b*) die because they are *Wnt3–*/*Wnt3–*. (*B*) Breeding scheme to balance a recessive embryonic lethal mutation. *Lim1* resides on chromosome 11 between *p53* and *Wnt3*. *Lim1–*/*Lim1–* mutants die around midgestation. To balance the *Lim1–* allele, *Lim1+*/*Lim1–* mice are first bred to mice carrying the chromosome 11 balancer chromosome (not shown). Agouti progeny with light ears and tails (i.e., carrying the balancer chromosome) are genotyped for the presence of *Lim1–*. Finally, *Lim–*/*b* males and females are intercrossed (shown). The only progeny that will survive to birth will be those with the same genotype as their parents (*Lim1–*/*b*). Therefore, no genotyping is necessary to maintain this stock. All you have to do is set up any of the progeny from these crosses as breeding pairs. When necessary, the *Lim1* mutation can be separated from the balancer chromosome by outcrossing to wild-type mice, genotyping the progeny, and selecting *Lim1* heterozygotes.

4.E.1 Complementation testing

Different alleles of the same gene can be very informative about gene function. Be watchful for other mutations, either spontaneous or induced in mutagenesis screens, for example, that resemble the phenotype of your mutation. It might be that the two genes are independent but both affect the same developmental pathway, resulting in similar phenotypes (see Section 4.E.3). Alternatively, it may be that they are both mutations in the same gene. This can easily be sorted out if both genes have been cloned, but if one or the other has not been cloned, you can test for allelism of the two mutations in question by what is known as a complementation test (Figure 4.5). This test is done by breeding heterozygotes for each mutation together. If the mutations are in separate genes, they will complement one another and the phenotype of all the offspring will be wild type, including the double heterozygotes (assuming both mutations are recessive). If they are alleles of the same gene, the offspring that inherit both alleles, i.e., the compound heterozygotes, have no wild-type allele and will have a phenotype resembling the homozygous phenotypes of the separate mutations. In other words, the mutations do not complement one another and we see a phenotype. The mutations will most likely differ molecularly because of their independent origins and therefore the compound heterozygotes may not be identical to either parental homozygous type. It certainly will be worth your while to determine the molecular lesion in the second allele to better understand the structural/functional relationship of the mutant proteins. With two mutant alleles and a wild-type allele, you are on the way to having an allelic series, which can be very informative about gene function.

4.E.2 Testing for a hypomorphic allele

Retaining a floxed or flrted selection cassette, or even a *loxP* site, within your targeted locus (e.g., within an intron in a conditional allele) might fortuitously produce a mutant effect. The promoter of the selection cassette may partially interfere with the targeted locus, or the cassette or the *loxP* site might have fortuitously disrupted a regulatory element. The allele may have wild-type activity, reduced activity, or no activity (a null). How do you determine which is the case? One way is simply to generate homozygotes for the allele to see if you get a mutant phenotype. If you already have a null allele for this gene, you can compare the two homozygous mutant phenotypes. If they are identical, then it is probable that the presence of the selection cassette or other exogenous sequence has disrupted the expression of the targeted gene to produce a null. If the phenotype is milder than the null phenotype, then you may have a hypomorphic allele. If so, then when you combine the potential hypomorphic allele with the null allele, the resulting phenotype should be more severe than the homozygous hypomorph, but less severe than the homozygous null. This can be very useful for understanding how different levels of gene activity influence development.

Another way to assess the activity of a targeted mutation is to combine it with a deficiency that includes the targeted locus. A number of chromosomes with relatively large (cM) deletions are available in mice that can be used for this purpose. Thus, if the phenotype of your targeted allele over the deficiency is less severe than the homozygous null mutant, then your allele may be a hypomorph. Some caution is necessary because the mutation over the deficiency will also be haploinsufficient for all of the other genes contained within the deficiency, which could influence the phenotype.

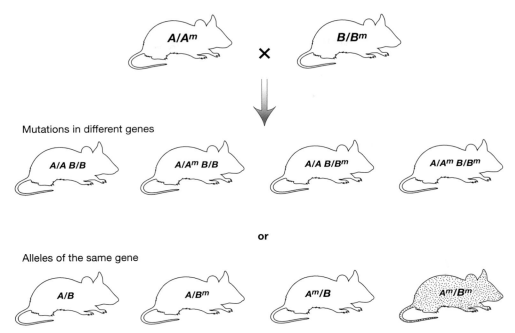

Figure 4.5. Test for allelism between two mutations. Suppose that you produced a mutation (A^m) in gene A and the homozygote has a similar phenotype to a mutation uncovered in an ethyl nitrosurea (ENU) screen. The ENU mutation was called B^m and although the gene has not yet been cloned, it maps to the same chromosomal region as A. To test if B^m is an allele of A, cross A/A^m mice together with B/B^m mice. If the mutations are in different genes, all of the offspring will be wild type in phenotype; if A^m and B^m are alleles of the same gene, one quarter of the offspring will be heterozygous for the two alleles (A^m/B^m) and will have a phenotype similar to A^m/A^m or B^m/B^m, depending on the nature of the molecular lesions.

4.E.3 Testing for genetic interactions

Mutations in many different genes may affect a given tissue. Combining heterozygous and/or homozygous mutations for different genes in the same mouse may reveal informative interactions between the genes. When the mutant phenotype of your gene of interest changes on different genetic backgrounds, it is the result of interactions between your mutation and genes that differ between the strains used to maintain the mutation (see Box 4.2). Epistasis describes a situation in which the effects of a mutation in one gene essentially mask the effects of a mutation in another gene, e.g., a null mutation in an enzyme that catalyzes an early step in a biochemical pathway will be epistatic over all genes that act on later parts of the pathway. On the other hand, a genetic interaction between two mutations would be indicated if mutations in different genes are brought together in a single mouse and the phenotype is different from the mutant phenotypes of the individual genes rather than simply a combination of the two. Genetic interaction is a broad term that does not suggest the mechanism of the interaction. The genes may function in either the same or parallel biochemical pathways. If they function in the same biochemical pathway, then the two gene products may have identical biochemical activities (functional redundancy) or they may be different components of the same biochemical pathway. If two genes function in parallel biochemical pathways, mutation of one gene may compromise the tissue and mutation of the second further compromises the tissue, producing a more severe abnormality because of the additive effects of both genes. Remember that a biochemical pathway can function within a cell or between cells. Thus, the expression patterns of the genes in question do not necessarily have to be coexpressed to justify a genetic interaction experiment.

It is very easy to breed two different mouse mutants together but the interpretation of the resulting phenotypes of the double mutants can be difficult. In addition, genetic interactions may show variable penetrance and variable expressivity so do not forget to do statistics (see Box 5.1, pp. 72–73). Below are some of the more common approaches that you can pursue when investigating a genetic interaction.

4.E.3.a Gene families: Redundancy or compensation

There is a reasonable chance that a gene knockout will produce a homozygous mutant mouse that is viable, fertile, and with no obvious abnormalities. Why does this occur so frequently? One explanation is that the mammalian genome has multiple copies of related genes that together regulate the same biological processes. Thus, if you have a mutation in only one member of a gene family, then you may not reveal an essential role for that particular gene unless you also include mutations in other family members. Certainly, if you are pursuing a gene-targeting experiment, search the databases to determine if a related gene(s) exists and if there is any overlap in its expression pattern with your gene. If there is no overlap, then the two genes will not have redundant functions. However, overlap of expression in a particular tissue during development or in the adult may result in tissue that does not show defects unless both genes are mutated because of functional redundancy.

Redundancy can be revealed by generating mice that carry heterozygous or homozygous mutations for each of the related genes (Figure 4.6A). Whether you discover new mutant phenotypes in the double heterozygotes (sometimes called transheterozygotes), in mice heterozygous for one gene and homozygous mutant for the other gene, or in double-homozygous mutants will depend on the total levels of the two gene products in a particular tissue at a specific time in development. It is relatively easy to document coexpression (e.g., by in situ hybridization or immunohistochemistry); however, it is difficult to measure the levels of gene products (mRNA or protein) in specific cells or tissues expressed in specific regions of a developing embryo or in adult tissues. If you can isolate the cells or tissue in question, then real-time reverse transcription–polymerase chain reaction (RT-PCR) or other quantitative assays of mRNA can be used; with sufficient amounts of tissue, a western blot can also be performed.

With respect to the effect on the phenotype, functional redundancy may be indistinguishable from a compensatory up-regulation, in which the expression of one gene family member is up-regulated in the absence of another, thus compensating for a single mutant effect (Figure 4.6B). As with functional redundancy, a phenotype or a more severe phenotype may only become evident when mutations in more than one family member are combined, but distinguishing redundancy from compensatory up-regulation requires knowing the level of gene expression of the putative compensating gene family member before and after combination with your mutation.

4.E.3.b Different components of a biochemical pathway

Many biochemical pathways that typically involve the participation of multiple unrelated proteins have been defined. Thus, you may already know that your gene encodes a component of a bio-

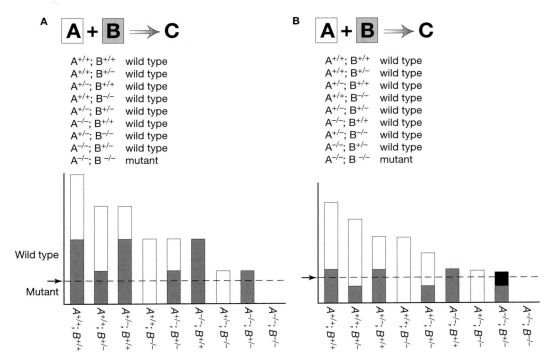

Figure 4.6. Redundancy and compensation. (*A*) Redundant action of gene family members coexpressed in the same tissue. A (*open boxes*) and B (*shaded boxes*) are gene family members that regulate a downstream target C. In this example, the wild-type level of the A protein is equal to the wild-type level of the B protein, and the bars on the histogram indicate the total level of A plus B proteins. The arrow and dotted line (*bottom*) indicate the threshold level of protein that is required in this situation to maintain a wild-type phenotype. When A and B levels are reduced by half (i.e., in the compound heterozygotes) or even by three-quarters (i.e., in the null, heterozygous genotypes), there are still sufficient total levels of A and/or B proteins to maintain a wild-type phenotype. Only when both A and B are eliminated in the double-homozygous genotype does a mutant phenotype develop. (*B*) Compensation by the up-regulation of a gene family member coexpressed in the same tissue. In this example, the wild-type level of the A protein is higher than the wild-type level of the B protein. The arrow and dotted line (*bottom*) indicate the threshold level of protein that is required in this situation to maintain a wild-type phenotype. One dose of A (i.e., $A^{+/-}$; $B^{-/-}$) is sufficient to maintain a wild-type phenotype, but one dose of B (i.e., $A^{-/-}$; $B^{+/-}$) would not appear to be sufficient. However, in the $A^{-/-}$; $B^{+/-}$ animals, the B locus is somehow sensitive to a deficiency in total A and B proteins and its expression is up-regulated sufficiently (*darker shading*) to result in a wild-type phenotype.

chemical pathway that has been defined in vitro. Are these in vitro findings also true in vivo? One way to test this possibility is to cross your mutant with mice mutant for different components of the biochemical pathway. Simplistically, you might expect the phenotypes of mutations in these genes to be similar, if not identical, if the mutations result in limiting amounts of gene products, and this has been seen many times (see Section 4.E.3.d) (Figure 4.7). However, it may be more complicated. An in vivo test to determine if the two gene products participate in a common biochemical pathway would be to cross the two mutants together to generate double heterozygotes. If a new phenotype appears, then you have evidence for a genetic interaction. Because you already had in vitro evidence that the two gene products participated in a biochemical pathway, it would be reasonable to suggest that this was also true in vivo. However, if the two components of the same biochemical pathway were not limiting in the single heterozygotes, then you might not see a new phenotype in the double heterozygotes, and the lack of a new phenotype in double heterozygotes would not preclude the possibility that the two genes functioned in the same biochemical pathway. You could obtain more proof that the two gene products truly acted in the same biochemical pathway if you could examine the expression of a downstream response gene.

4.E.3.c Different genes with overlapping expression patterns

Expression profiling and sequencing of expressed sequence tags (ESTs) from cDNA libraries have yielded tremendous amounts of information about gene expression. Thus, if you are studying a particular tissue, you will probably know a lot about the expression of other genes in that tissue. Take advantage of this information and the availability of many mouse mutants. Cross your mutant with other mice that have mutations in genes coexpressed in the tissue in which your gene is expressed to look for new mutant phenotypes in double mutants. Be cautious about interpretations by keeping in mind the types of proteins encoded by the two different genes. If no biochemical connection is obvious or you were unable to establish such a connection, then you may simply be compromising the function of a specific tissue in different ways by the additive effects of each mutation, not necessarily revealing a common biochemical pathway. For example, one mutation may slightly alter a tissue without causing an obvious phenotype. A mutation in a different gene may have the same effect. However, when both mutations are brought together, the two subthreshold defects combine to alter the tissue (e.g., proliferation, migration, adhesion, etc.) enough to produce a mutant phenotype.

Once you establish a genetic interaction, more biochemical information will be needed to formulate a molecular/biochemical interpretation. However, even if you do not find that the two gene products act in the same biochemical pathway but are probably acting in parallel pathways, the double mutants can be useful since the genetic interaction still results in a mutant tissue. Thus, these tissue-specific alterations can be helpful for understanding the role of a specific tissue during embryonic development or adult tissue function.

4.E.3.d Similar mutant phenotypes caused by unrelated genes

Once you know the mutant phenotype of your gene, look for other mutations that result in the same or similar phenotypes. Mice with identical mutant phenotypes suggests that the gene products may participate in the same biochemical pathway (e.g., ligand and receptor or two transcription factors that form a complex) (Figure 4.7). Thus, the two genes may be coexpressed in the same cell but could also be expressed in different cells. First, determine the cells and tissues in which the two genes are normally expressed. Then, check the expression of your gene in mice mutant for the other gene and vice versa to investigate a possible epistatic relationship. Generate double heterozygotes. Do the double heterozygotes have a phenotype that is not found in the individual heterozygotes? If they do, then you have evidence for a genetic interaction. At this point, you will have to do more experiments to determine if there is a biochemical relationship between the two gene products. As pointed out in Section 4.E.3.b, a negative result would not provide enough information to draw conclusions.

4.E.3.e Mutant phenotypes caused by overexpression of downstream genes

A mutation may eliminate a negative regulator of a downstream gene or gene product (i.e., an antagonist). Thus, a mutant phenotype may actually be caused by the overexpression of a down-

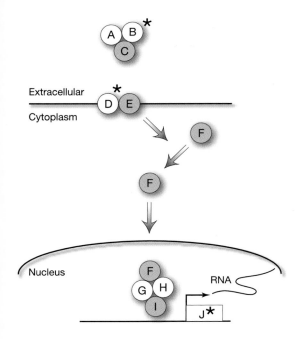

Figure 4.7. Genes in the same biochemical pathway. Extracellular proteins (A–C) can interact with cell membrane proteins (D and E) to transduce signals, perhaps by modifying cytoplasmic proteins (F) that in this example translocate into the nucleus to form a transcriptional complex (F–I) leading to the expression of a downstream target (J). In vitro studies elucidating this pathway provide a framework for in vivo genetic analyses. Mutations (*asterisks*) that lead to alterations in individual components of the pathway could compromise the pathway leading to a mutant phenotype in animals. Combining mutations for multiple components of a biochemical pathway (i.e., generating double or triple heterozygotes) may reveal a genetic interaction if the gene products of the individual components are at limiting amounts when heterozygous.

stream gene (Figure 4.8). This may be indicated by a molecular marker analysis if you find that one of the markers is up-regulated in a particular tissue that is abnormal. According to this hypothesis, if you reduced the expression of the downstream gene in the mutants, you should suppress the mutant phenotype, indicating a genetic interaction. Therefore, if you suspect such a situation, you can test this idea by crossing your mutant with mice carrying a mutation in the downstream gene, thus reducing the number of wild-type alleles of the downstream gene. This should reduce the amount of downstream over expression and suppress the mutant phenotype.

4.E.4 Increasing the frequency of homozygous mutant mice

Germ-cell-expressing Cre mouse lines can be used in combination with conditional alleles to increase the frequency of homozygous mutant mice from 25% to up to 100%, which could be very useful for analyses such as expression profiling using DNA microarrays. This strategy has been dubbed "cheating Mendel" because the outcome gives the illusion that Mendel's laws

Model   A ⊣ B

A ⇥ B↑

Result

Genotype	Phenotype
$A^{+/+}$; $B^{+/+}$	wild type
$A^{-/-}$; $B^{+/+}$	mutant phenotype
$A^{-/-}$; $B^{+/-}$	suppression of mutant phenotype

Figure 4.8. Negative regulator of a downstream gene. If A encodes a negative regulator of B, then a phenotype resulting from the loss of A could be due to the up-regulation (overexpression) of B. This model could be tested by reducing the number of wild-type copies of B (e.g., heterozygosity of B). Suppression of the A mutant phenotype would provide genetic evidence for the model.

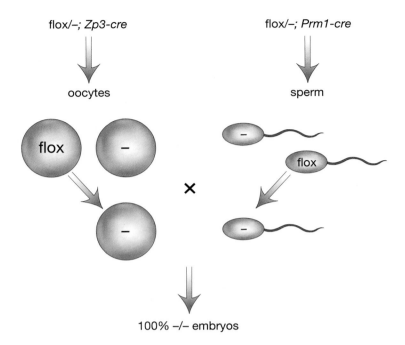

Figure 4.9. Cheating Mendel. Germ-cell-specific recombinases can be used in combination with floxed (or flrted) alleles to increase the proportion of homozygous mutant mice or embryos in a litter. For example, *Zp3-cre* female mice express Cre in oocytes and *Prm1-cre* male mice express Cre in spermatids (*shading*). Females can be generated that are heterozygous or homozygous for the floxed allele and hemizygous for the oocyte-specific *cre* transgene (flox/–; *Zp3-cre* or flox/flox; *Zp3-cre*) and males can be generated that are heterozygous or homozygous for the floxed allele and hemizygous for the spermatic-specific *cre* transgene (flox/–; *Prm1-cre* or flox/flox; *Prm1-cre*). Cre expression in the oocytes of females carrying *Zp3-cre* will convert floxed alleles to null (–) alleles. Likewise, Cre expression in the spermatids of males carrying *Prm1-cre* will convert floxed alleles to null alleles. Thus, females produce only null oocytes and males produce only null sperm. From this cross, only homozygous mutant (–/–) mice or embryos can be generated.

have been broken (Figure 4.9). Here's how it works: First, obtain transgenic mouse lines that express Cre in the male and female germ lines. For example, protamine 1 (*Prm1*)-*cre* transgenic mice express Cre in spermatids and zona pellucida 3 (*Zp3*)-*cre*-transgenic mice express Cre in oocytes. Breed these *cre* transgenes to your conditional allele. Ultimately, you will end up with mice that are homozygous for the conditional allele and hemizygous for the germ-cell-specific *cre* transgenes. Thus, when the male generates spermatids, Cre expression will convert them from conditional to null. This is facilitated by the syncytial nature of spermatogenesis in which gene products from one nucleus are shared with neighboring sperm cells through a common cytoplasm. In females, the developing oocytes will express Cre before the final meiotic division, loading the oocytes with Cre protein. Thus, the conditional allele will be converted to a null in the resulting oocytes. When these males and females are bred together, they can only generate homozygous recombined (null) embryos. In practice, the yield of homozygous recombined embryos may be a bit lower than 100%, depending on the efficiency of germ cell Cre activity. You can also manipulate this breeding strategy to generate 50% homozygous mutants if one of the parents is heterozygous for the conditional allele. This system is especially useful in the rare cases of a haploinsufficiency lethal mutation.

CHAPTER 5

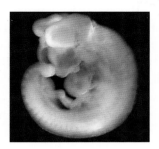

Phenotypic Analysis: Prenatal Lethality

THIS IS THE STARTING POINT FOR THE ANALYSIS of spontaneous or induced mutations in heterozygous or homozygous mutant mice. How ever you reached this chapter, you now have the long-sought phenotype resulting from either a targeted mutation in your favorite gene, a mutation in an unknown gene, or a selected mutation of your choice. The phenotype may be just what you predicted based on gene expression and gene product characteristics, or, if it is a spontaneous mutation, you might have selected it because of the interesting phenotype. But quite possibly, the mutation resulted in a different phenotype than you expected, or novel mutant characteristics in addition to those you expected. At first analysis, it may even look like there is no mutant phenotype at all, but this is unlikely, and a closer examination or some manipulations will usually reveal the effects of the mutation.

This chapter presents strategies to pinpoint the timing of the mutant effect and analyze a prenatal lethal mutation. It is written for the situation in which the mutation is recessive and thus the phenotype is being analyzed in the progeny of heterozygous mice that are viable and fertile. However, the analytical methods are applicable to dominant effects as well (Chapter 7). Thus, we begin by examining the progeny of matings between heterozygous mice.

5.A   Are Homozygous Mutants Present at Birth?

The first question is whether the homozygous mutants are present at birth...but how will you know if they are or not? It is impractical to genotype mice at birth, so most likely the first time

---

Chapter opening artwork courtesy of Zhou-Feng Chen and Richard Behringer.

you will be aware of missing homozygous mutants is when you genotype the mice between 10 days and weaning. If they are not present then, they could still be born but die sometime between birth and genotyping. As a routine, check females near the expected time of parturition, count newborn offspring, and look for any dead pups (which you can genotype). At the time of genotyping and weaning, see if the number of pups correlates with the number born, or whether postnatal losses and the lack of homozygous mutants indicate that they were born but died before weaning, in which case, you should go to Chapter 6.

Homozygous mutants are present at birth	Go to Chapter 6 ▶
Homozygous mutants are not present at birth	See Possible Causes ▼

Possible Causes

a. *First trivial possibility.* The number of progeny analyzed, i.e., the sample size, might be too small to be statistically significant. Do not jump to conclusions based on small numbers. Instead, review Mendelian ratios and do a Chi square (χ^2) test of significance (Box 5.1). A low probability is still a probability, and you are the one who decides the significance level. Breed more mice until you are convinced that no homozygous mutants are born, and check through the other possible causes listed here.

Box 5.1 **Worked Examples of a Chi Square Test of Mendelian Segregation**

If you want to know whether the frequency of genotypes you observe is what you expect on the basis of Mendelian segregation, a χ^2 test is appropriate. It goes without saying that the larger your sample, the more confidence you can have in the result. The test provides a probability only; you will have to decide the level of significance, although 5% is typically used. After performing the test, you will also see where the discrepancy lies between what you observe and what you expect.

First example

Let us say that you have collected 36 postimplantation embryos from a heterozygote × heterozygote cross, 34 of which you successfully genotyped (the polymerase chain reaction [PCR] for the other two failed), and you found the following distribution of genotypes: 10+/+, 22+/–, and 2–/–. In addition, during the dissections, you noted ten decidual swellings that contained only trophoblast giant cells and could not be genotyped. From the 34 embryos genotyped, you would expect the following distribution of genotypes under Mendel's laws: 8.5+/+, 17+/–, and 8.5–/–, which is the perfect 1:2:1 ratio. Obviously, we are not dealing in half embryos, so this is only the theoretical expectation. Now, compare your results with what you expected using the following formula, which provides a measure of the deviation of each observed value from the expected value for each genotypic class:

$$\frac{(\text{observed} - \text{expected})^2}{\text{expected}}$$

Add these deviations for each genotypic class and you have the Chi square statistic (χ^2):

$$\frac{(10-8.5)^2}{8.5} + \frac{(22-17)^2}{17} + \frac{(2-8.5)^2}{8.5}$$

$$0.26 \quad + \quad 1.47 \quad + \quad 4.97 \quad = \quad 6.7 = \chi^2$$

Any elementary statistics text contains tables of this statistic. Using the appropriate degree of freedom, locate this χ^2 value in the table and read off the probability. The degree of freedom is the number of categories that can vary independently given the number of observations. In other words, with 34 embryos, two categories could vary independently, but the third will then be fixed as the difference between 34 and the sum of the numbers in the other two classes. In this case, with three genotypic categories, there are two degrees of freedom. The χ^2 value corresponds to a probability of between 5% and 2%. This means that the probability of getting this distribution of genotypes out of 34 embryos by chance is less than 5% (but greater than 2%). If you set the level of significance at 5%, the distribution is significantly different from expected, but if you decide to go with a more stringent level of significance, say, 1%, then it is not significantly different from expected at the 1% level. As we said, the 5% level of significance is the usual, but being so close to the significance level means that you might want to collect a larger sample before drawing final conclusions.

An examination of the individual values contributing to the χ^2 value can also be informative. This example has close to the expected number of +/+ embryos and so the deviation is small, as is the contribution to the χ^2 value. The number of –/– embryos is quite far off from the expected and the deviation is correspondingly large, making a relatively large contribution to the χ^2 value, and indicating that embryos are missing from this class. However, the deficiency in +/– embryos also contributes a fairly substantial proportion to the total deviation.

Finally, think about those ten decidua containing giant cells and the two embryos that were excluded from the calculations because they were not successfully genotyped. To the extent that technical losses, such as the failure of a PCR reaction, is independent of the genotype of the embryo, we are probably safe to exclude the two lost embryos from the calculations. However, biological losses as represented by the "embryos" that consisted only of giant cells could indeed be genotype dependent. In fact, this could be a phenotype. For the sake of argument, you might want to include these giant-cell-only deciduas in the –/– class and do another χ^2 test, under the assumption that these are mutant embryos. Now does your distribution fit Mendel's prediction?

Second example

You are genotyping offspring of a heterozygous cross at the time of weaning. You do not observe any homozygous mutant offspring and want to know how many you should genotype before concluding that the homozygous mutant animals die sometime before weaning. If you observe a perfect 1:2 ratio of +/+ to +/– animals with no –/–, then by using the χ^2 test, you reach significance at the 5% level with 18 offspring, and at the 1% level with 30 offspring, and you can be 95% or 99% confident, respectively, that the homozygous mutant animals die before weaning. However, these levels of significance will be reached with even fewer offspring if there is a deficiency of heterozygotes, because this will also contribute to the χ^2 value. If you see a deficiency of +/– animals, it may mean that something additional is operating, such as a heterozygous effect on viability.

b. *Second trivial possibility.* One of the parents might have been incorrectly genotyped. Do you find a 2:1 ratio of heterozygous to wild-type offspring that might indicate missing homozygotes from a heterozygous cross? Or do you find a 1:1 ratio that might indicate a misgenotyped parent, resulting in a cross between a wild type and heterozygote and thus a 1:1 heterozygous to wild-type ratio in the offspring? If this is the case, re-genotype the parents or mate different animals. If not, continue checking the other possibilities.

c. *Third trivial possibility.* You are experiencing technical problems with the genotyping assay. Up to this point, you have only needed to distinguish wild-type and heterozygous mice. Double check that your PCR or Southern assay can distinguish the homozygous mutants as well and then check every technical aspect of the assay to make sure it is working correctly.

d. The *nontrivial possibility* is that the mutation results in prenatal lethality. Go to Section 5.B.

5.B Assessment of Time of Death of Homozygous Mutants During Embryonic Development

Narrowing down the time of death of mutant embryos is the first step toward analyzing the cause of death and can be accomplished with very few litters. Of course if you are challenged with a phenotype displaying a variable time of death, it may take a while longer, a few more litters, and a lot of patience to pin down the range of time of death. In either case, the idea is to pick several time points in gestation (Figure 5.1) at which embryos can be easily genotyped and a morphological assessment of the embryonic and extraembryonic structures can be done easily to provide a maximum amount of information about phenotype. Do not make the mistake of trying to cover every developmental stage in this initial screen. Concentrate your efforts on a few stages to be confident of your observations and get enough data to do statistics. These dissections serve a dual function: First, dead or abnormal embryos may be discovered that, on genotyping, are found to be the homozygous mutants, and thus the dissections provide material for analyzing the mutant phenotype. Second, the time of death and/or the time of onset of the abnormalities can be inferred from the developmental stage and condition of the mutant embryos, or alternatively, the absence of mutant embryos among those recovered can indicate very early lethality. Below we describe the way in which the system works. You can modify the details of these four steps to adjust them to your needs.

1. Start by sacrificing 1–3 pregnant heterozygous females, mated to heterozygous males, at 12.5 days postcoitus (dpc). This stage is chosen because the embryos are easy to dissect and

HELPFUL HINT

Once a stud male genotyped as heterozygous has been proven heterozygous by the production of a litter containing homozygous mutants, he can be marked as a proven heterozygote and used with confidence. Females genotyped as heterozygous should always be double checked at the time of their use in case of an error in the original genotyping.

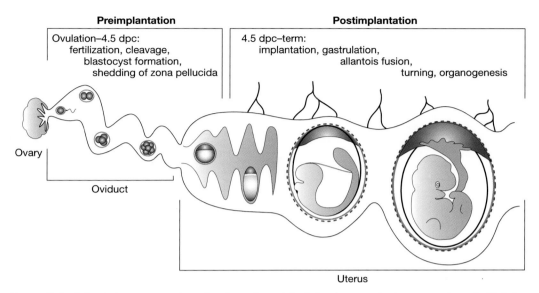

Figure 5.1. Schematic representation of embryonic and fetal development in the mouse through 12.5 dpc. The period of development before implantation in the uterus (preimplantation development) is characterized by fertilization, cleavage, compaction, blastocyst formation, and hatching from the zona pellucida. Embryos move through the oviduct and traverse the uterotubal junction (between the uterus and the oviduct) at about 2.5 dpc. Attachment and implantation into the uterus follow at 4.5 dpc and, during postimplantation development, the embryos undergo gastrulation and organogenesis. Attachment of the allantois to the chorion to form the chorio-allantoic placenta is an important developmental landmark at about 8.5 dpc. Between 8.5 and 9.5 dpc, embryos turn from a dorsal flexure to a ventral flexure or fetal position. The period between 12.5 dpc and term is characterized by organogenesis and growth. (Modified from Papaioannou and Hadjantonakis, 2003, *Stem cell handbook* [ed. S. Sell], pp. 19–31, Humana Press, Totowa, New Jersey.)

because it provides a lot of information about what has been going on during the first half of gestation.

Take a tissue sample from each mother to reconfirm the genotype (the kidney yields very nice DNA for Southern analysis or PCR). Count any swellings of the uterus, which represent the sites of embryo implantation (Figure 5.2), no matter what their size. Then, remove the ovaries, keeping track of right and left so that they can be correlated with the right or left uterine horns. Have someone else make counts of the corpora lutea (CL), or do it blind (see Box 3.3, p. 33). The CL counts correspond to the number of oocytes ovulated and therefore to the number of implantation sites plus any preimplantation losses, which will include failure of fertilization as well as embryonic death. Because mouse embryos almost never cross from one uterine horn to the other, the correlations between CL counts and implantation sites should be made on the basis of each uterine horn. Of course, you could wait until after you genotype the implanted embryos and discover that the homozygous mutant embryos are not among them, but CL data is very simple to collect and this foresight could save you several precious heterozygous females and a few days or weeks of time later.

2. Next, examine each fetus and implantation site by careful dissection (for an excellent reference with good pictures and instructions, use the mouser's bible, *Manipulating the Mouse*

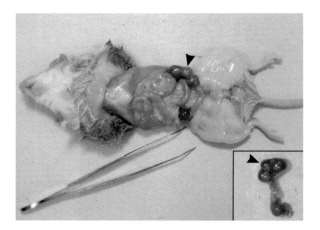

Figure 5.2. Implantation sites at 12.5 dpc. The abdomen has been opened and the fat pad reflected caudally to reveal four implantation sites, one in the right uterine horn and three in the left. The middle site (*arrowhead*) contains an abnormal embryo and is smaller than the others. (*Inset*) The uterus after dissection from the female.

Embryo [MM3*]). You will want to collect some fetal tissue for genotyping and also to assess the embryo, placenta, and fetal membranes for normality, so a systematic dissection—first removing maternal tissue and then progressively working through the fetal membranes to get to the embryo itself—is called for. This will give you a chance to evaluate each conceptus thoroughly. Separate the embryos into individual dishes or drops of media, give them numbers right away, and take notes as you go. Make use of the various resources listed in Appendix 1 to learn embryo dissection and anatomy, but even without a lot of experience, careful observation and comparisons could turn up differences among embryos. Photograph anything that looks out of the ordinary. The subsequent genotyping of embryos will then reveal the significance of these differences.

3. At this point, collect embryonic tissue for further analysis. You will certainly want to do this if you see any potential abnormal phenotypes. It is very little work to preserve each numbered embryo and its fetal membranes separately in small vials or tubes, and they can be discarded later if they are not needed. The fixative of choice will depend on the most likely analysis to be done (see later in this chapter). For genotyping, it is simplest to use the entire yolk sac, which will provide enough DNA for one Southern blot or many PCR reactions, but be careful to separate it completely from the placenta to avoid maternal contamination. An alternative source of tissue for genotyping, especially if you think the yolk sac might have a phenotype that you want to investigate, is a leg of the embryo, leaving the other three legs as well as the yolk sac for future analysis. Whatever embryonic sample you take for genotyping, be sure to wash it extensively to avoid maternal contamination because the mother is, or should be, a heterozygote.

4. Once you have genotyped the fetal (and maternal) samples, you are very close to pinning down the time of embryonic lethality. If you have not recovered any homozygous mutant

*MM3 refers to *Manipulating the mouse embryo: A laboratory manual*, 3rd edition (Nagy et al. [2003]. Cold Spring Harbor Laboratory Press, Cold Spring Harbor, New York).

embryos among the implanted 12.5-dpc embryos or degenerating remains, and the CL counts are at least 25% higher than the number of implantation sites, the mutant embryos probably died between fertilization and implantation at 4.5 dpc, leading you to Section 5.C. Do not be fooled by small numbers, however. Before you jump to this conclusion, double check the mothers' genotypes and gather enough data to get a respectable p value in a χ^2 test (see Box 5.1). Once you are sure that the homozygous mutants have been lost before implantation, go to Section 5.C. Otherwise, carry on.

If the embryos died between implantation at 4.5 dpc and about 9.5 dpc, before the establishment of the allantoic circulation, they will be represented by implantation sites that contain only degenerating remains. These sites are variously referred to as empty decidua, moles, or resorption sites. They may actually be empty or contain remnants of extraembryonic structures: trophoblast giant cells, yolk sac, Reichert's membrane, and/or the ectoplacental cone. In the absence of an allantoic circulation, however, very little in the way of embryonic (as opposed to extraembryonic) remains will be recognizable by 12.5 dpc, because the embryo will be in an advanced state of degeneration. Obviously, the number of these resorption sites should match the number of "missing" homozygous mutants if that is what they represent, but sufficient remains may allow you to genotype them for a positive identification. Go to Section 5.D for early postimplantation phenotypes.

If embryonic death occurred after the establishment of the allantoic circulation, or has not yet occurred by 12.5 dpc, the mutant embryos will either be alive, as defined by the presence of a beating heart, or present in varying stages of degeneration depending on when they died. If embryos have no heartbeat, their developmental stage can be used to estimate the time of death (Box 5.2). Degenerating embryos can sometimes be genotyped if the DNA has not degraded too much, but otherwise, their number should match the number of "missing" homozygous mutants. Go to Section 5.E for midgestation to late-gestation phenotypes.

If all of the embryos including homozygous mutants are viable at 12.5 dpc, embryonic lethality must be occurring between 12.5 dpc and term. Detailed examination of the homozygous mutants, numbered and preserved along with the littermate controls, should be done at this point, because the onset of the mutant effect may already be evident. Dissection of embryos at a later time point near term, say, 16.5 dpc, will pin down the time of death.

Finally, keep in mind that there may be variable penetrance of the phenotype and variable expressivity, especially if the embryos have a mixed genetic background. Be sure to look at enough embryos to determine if the homozygous mutant phenotype is uniform and the mutants are all dying at the same time. Again, concentrate on this single stage to get sufficient numbers to be confident of your conclusion and then go to Section 5.E.

5.C Analyzing Preimplantation Lethal Phenotypes

A preimplantation lethal phenotype is indicated if no homozygous mutant offspring are born, no homozygous mutants are detected among implanted embryos, and there is not an excess of empty or abnormal implantation sites. There should also be a 25% excess of CL over the number of implantation sites, by comparison with the number in a control backcross (heterozygous × wild type) to take normal variation into account (Box 5.3). For all these determinations, col-

> **Box 5.2** Useful Developmental Landmarks for Embryos Between 4.5 dpc and Term
>
> You may want to determine the developmental stage of an embryo if you need to estimate gestational age or the time of death of an embryo, or determine if a mutation causes a developmental delay compared with wild-type littermates. Developmental rates and the length of gestation vary somewhat in different strains of mice, and some developmental asynchrony is often found within litters. However, certain easily scored structures provide landmarks for comparing developmental stages within or between litters and/or estimating chronological age of a litter if this is unknown. The day of the plug is called 0.5 dpc (or E0.5, for embryonic day 0.5), making the assumption that fertilization took place around midnight, although this may not be a justified assumption if the mice were together all night. If you believe that you missed detecting a vaginal plug, you can look for the appearance of blood in the vagina of the mother, which occurs at 10.5 dpc, when the uterine lumen reopens. Otherwise, the following pointers can help in estimating embryonic age and/or staging embryos of known gestational age:
>
> - Implantation takes place at 4.5 dpc.
> - 5.5–8.5 dpc. Developmental stage at these ages is quite variable even within litters because of the rapid rate of growth and development, but the presence or absence of a circumferential embryonic-extraembryonic constriction, as well as the overall size of embryos, is easily scored. For detailed staging of embryos during the latter part of this period, see Box 5.10.
> - Fusion of the allantois to the chorion takes place around 8.5 dpc when the embryo has 5–8 somite pairs.
> - Embryos turn from a dorsal curvature (lordosis) to a ventral curvature (fetal position) by 9.5 dpc (between 6 and 16 somite pairs) (Figure 5.3A,B).
> - 8.5–12.5 dpc. The number of somites can be easily counted for finely graded developmental staging (see Box 5.11).
> - Neural tube closure begins at around 8.5 dpc at the 6- to 7-somite stage and is completed by 10.5 dpc at the 29- to 30-somite stage (see Figure 5.17).
> - 9–18.5 dpc. The shape of the limb buds reflects developmental stage: Forelimb buds first appear as lateral ridges at the level of somites 7–12 at 9 dpc and hindlimb buds appear a day later at the level of somites 23–28. This asynchrony in first appearance of the forelimbs and hindlimbs is maintained in their later development.
> - 10.5 dpc. Eye pigment appears, unless, of course, the embryo is albino.
> - 10.5 dpc. A loop of the midgut herniates into the umbilical cord (Figure 5.3C). This is known as a physiological umbilical hernia and it persists until the gut loop is withdrawn into the peritoneal cavity at 15.5 dpc.
> - Features of the head including the branchial arches, the mouth and nose region, the eyes, and the external ears all follow a strict developmental progression and can be used to stage and compare later embryos (see Section 5.E).

lect sufficient data to provide statistical support: χ^2 analysis (Box 5.1) using the backcross frequencies to estimate a corrected expected Mendelian frequency will be most accurate.

Once you are sure that the homozygous mutant embryos are dying before they implant, you have narrowed the time window to about 4.5 days—the time between fertilization and implanta-

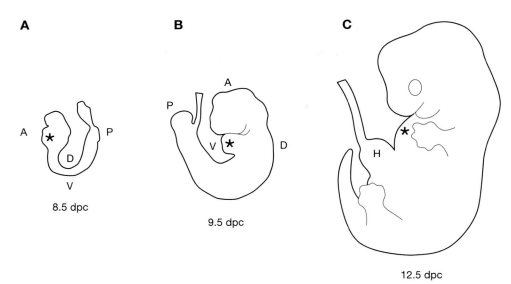

Figure 5.3. (*A*) Lordosis of an 8.5-dpc embryo before turning. The embryo has a dorsal curvature. (*B*) Fetal position of a 9.5-dpc embryo after turning. Embryos in the fetal position have a ventral curvature. (*C*) Umbilical hernia (H) at 12.5 dpc. Between 10.5 and 15.5 dpc, a loop of midgut is located outside the abdominal cavity in the base of the umbilical cord. The placenta and yolk sac are not shown. (A) Anterior; (P) posterior; (V) ventral; (D) dorsal; (*) location of the heart.

Box 5.3 Establishing Control Levels of Embryonic and Fetal Loss

The number of embryonic and fetal losses that occur for reasons unrelated to a specific mutant genotype, i.e., the control frequency of embryonic loss, varies with different strains of mice and must be determined empirically. The best way to do this is to set up a backcross (heterozygote × wild type) by mating a heterozygous male with wild-type females and vice versa, and then dissecting embryos at the gestational stage you want to compare to count the dead, dying, or missing embryos. In this control cross, the genotypes of the embryos will match the genotypes of embryos from an intercross (heterozygote × heterozygote), except for the absence of the homozygous mutants, and thus the control frequency of embryonic death can be used to calculate a modified, expected Mendelian ratio for the intercross. For example, if the control rate of embryo recovery from a backcross is 92 embryos/100 CL detected, then the expected recovery from the intercross, assuming that the mutant effect is recessive and the homozygous mutant embryos are not recovered, is 25% less than this figure, i.e., 69 embryos/100 CL. This takes into account the 8% random loss (8 embryos) found in the backcross and a 25% loss of the remainder due to the homozygous mutant genotype (23 additional embryos).

When determining the control level of embryonic loss, be on the lookout for dominant effects that could affect the heterozygous embryos or the reproductive capability of the heterozygous mothers. Major dominant effects on embryonic development should have been detected during earlier analyses, but heterozygous effects on reproductive capacity might have been missed if they are subtle. To control for this possibility, use reciprocal matings of either heterozygous males or heterozygous females × wild type when determining the control level of embryonic loss.

tion. But it is still important to find the earliest time point at which the mutant departs from normal to determine the kind of effect the mutation is having. During this preimplantation period (Figure 5.4), the zygote should go through six or seven special mitotic divisions called cleavage (so called because no overall growth of the embryo occurs during this process, but simply a cleaving of the cytoplasm accompanied by nuclear division), resulting in smaller cells called blastomeres. Toward the end of cleavage, the blastomeres compact into a tight spherical ball of cells known as a compact morula (Figure 5.4F), and cellular differentiation begins with the formation of an outer layer of trophectoderm and an inner cell mass (ICM) (Figure 5.4G). The trophectoderm begins to pump fluid, and the process of cavitation results in a cystic ball of cells, the blastocyst, with the ICM located asymmetrically at one pole. The trophectoderm overlying the ICM is known as the polar trophectoderm, and this is the embryonic pole of the embryo; the trophectoderm on the opposite or abembryonic pole is called the mural trophectoderm. The cavity is called the blastocyst cavity (or blastocoel). Next, a third cell type, the primitive endoderm, forms on the blastocoelic surface of the ICM. All of these events take place on a fairly precise time schedule, although there will be some minor asynchrony, and a morphological assessment is straightforward.

Examination of embryos at intervals after fertilization will quickly tell you whether some embryos either lag behind or arrest their development. The object is to identify the point at which embryos first begin to stray from the straight and narrow path and then to genotype them to confirm that this phenotype correlates with the mutant genotype. Theoretically, even a single cell from a preimplantation embryo can be genotyped with a sensitive PCR assay, but, depending on the time of death or arrest, you might have as many as 100 cells with which to work. Keep in mind that the zygotic genome in the mouse begins to function during as early as the two-cell stage, so a mutant defect might become evident very early if the gene in question has a role in some basic cellular functions such as cell division or energy metabolism.

Start the examination at 3.5 dpc, when the embryos can be flushed from the uterus as expanded blastocysts. The zona pellucida (ZP), a transparent, extracellular envelope that is deposited around oocytes during oogenesis, will still be intact around the embryo and any embryos that arrested or died should be recoverable as ZP-contained remains. It is a good idea to count CL (see Box 3.2, p. 31) to make sure that all of the ovulation products are recovered (this is especially important if for some technical reason there is no means of genotyping the preimplantation embryos, in which case, conclusions will be based on expected Mendelian frequencies). Mice usually have a very low rate of preimplantation failure, although this may vary with the strain and it might be worthwhile to analyze a backcross to establish the control level of preimplantation failure (Box 5.3). In fact, it is essential to know the control level of preimplantation failure if the embryos cannot be genotyped. To make an assessment of preimplantation development, sort the embryos into the following morphological categories (Figure 5.4): degenerated or fragmented; unfertilized oocytes; zygotes; two cell, etc., up to eight cell; compact morulae; cavitating blastocysts; and expanded blastocysts. Then, genotype the embryos and correlate their genotype with their phenotype. If the homozygous mutants are among the normal-looking blastocysts, go to Section 5.C.2. If they are either retarded, abnormal, or among the degenerating embryos, continue to Section 5.C.1.

5.C.1 Techniques for the analysis of cleavage stages and early blastocysts

Whether the mutant embryos are degenerating, arrested at a particular stage, or have abnormal morphology at the blastocyst stage, the object is to gain a dynamic picture of how they got that way. Fortunately, preimplantation embryos are very easy to culture from fertilization right

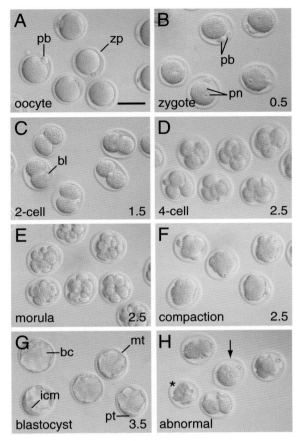

Figure 5.4. Preimplantation development. Differential interference contrast (DIC) images with days postcoitus indicated. (*A*) Ovulated oocytes have a single polar body and a single haploid nucleus. (*B*) Zygotes are easy to distinguish from unfertilized oocytes by the presence of two polar bodies and two pronuclei and also by the shrinkage of the zygote, resulting in a larger subzona space. (*C–E*) Two-cell to eight-cell cleavage stages. (*F*) Compact morulae. The individual blastomeres are no longer distinguishable due to tighter cell–cell adhesion that occurs during compaction. (*G*) Blastocyst-stage embryos consisting of an outer layer of trophoblast cells surrounding an inner cavity, the blastocyst cavity, with an inner cell mass (ICM) located at one pole. The trophoblast overlying the ICM is the polar trophoblast, whereas the rest of the trophoblast is the mural trophoblast. (*H*) Abnormal embryos that are fragmenting (*), shrunken (arrow), and degenerate. Ovulated oocyte to morula stages are found in the oviduct, whereas blastocysts are found in the uterus. Ovulated oocytes to blastocysts are surrounded by a zona pellucida. Blastocysts expand and subsequently "hatch" out of the zona pellucida to implant into the uterus. Scale bar, 100 μm. (bc) Blastocyst cavity; (bl) blastomere; (icm) inner cell mass; (mt) mural trophoblast; (pb) polar body; (pn) pronucleus; (pt) polar trophoblast; (zp) zona pellucida.

through cavitation to the blastocyst stage and hatching from the zona pellucida (Box 5.4), and watching a clutch of embryos progress through cleavage is a satisfying way of simply waiting for the mutants to reveal themselves. Collect the embryos from the oviducts at 0.5 dpc (MM3), sort them according to the presence or absence of two pronuclei (Figure 5.4B), set aside any unfertilized oocytes for further observation, and culture them in case fertilization has simply been delayed. Then culture the zygotes individually but identified by litter (this way you can check the genotype of the mother in the interim and discard any litters from mothers who were misgenotyped). Score the embryos daily with a good dissecting microscope or with phase or

DIC microscopy, using the classifications in Section 5.C. At the end of 4 days of culture (or less, if the mutant phenotype is evident earlier), genotype each embryo and correlate genotype with phenotype to confirm the mutant status of abnormal embryos. As always, if something technical prevents you from genotyping the embryos, you will need to know the normal variation in a control backcross to make a valid statistical comparison (see Box 5.3).

Cleavage and the formation of a normal blastocyst involve several related, but not necessarily dependent, processes. First, cell number increases and cell size decreases with time. A genetic program for development begins to unfold from the zygotic genome starting at the two-cell stage through the onset of expression of specific genes. Concomitantly, cellular interactions take place among the blastomeres that influence the spatial distribution of cytoplasmic organelles and

Box 5.4 Hints on Preimplantation Embryo Culture to Detect a Mutant Phenotype

In vitro culture of preimplantation mouse embryos is an excellent means of observing events that normally take place in the mother's reproductive tract. Superovulation of female mice can be used to increase the number of embryos obtained per female (MM3). Protocols for embryo collection and culture are fairly standard, and detailed protocols for embryo isolation techniques and culture media are found in the mouse manual (MM3). Preimplantation embryos are quite robust and will survive a range of culture conditions; however, their development is compromised in suboptimal conditions. To recapitulate in vivo development as closely as possible so that a mutant phenotype can be distinguished, rigorous attention must be paid to the details of culture, particularly the culture media, pH, and temperature. Although several popular media formulations have been in use for decades, some of these clearly do not support optimal development and should be avoided. KSOM + AA is the medium of choice for the culture of zygote through blastocyst stages, with incubation in 5% CO_2 in air or 5% CO_2/5% O_2/90% N_2. For flushing and handling embryos before culturing, flushing and handling medium (FHM), which contains HEPES and reduced bicarbonate, is a good choice to maintain the correct pH in air. As embryos develop, their culture requirements change. For attachment and further growth of blastocysts in vitro, it is necessary to have serum in the medium and culture the embryos on tissue culture plastic (as opposed to bacteriological petri dishes). A good choice of medium for growing blastocysts over the periimplantation period, as well as for culturing isolated ICMs, is embryonic stem cell medium with 10% fetal calf serum for incubation in 5–7% CO_2 in air (MM3).

When comparing embryos of different genotypes, each embryo must be cultured separately for identification purposes. This is easily done by setting up a series of numbered drops of medium (10–20 μl) in plastic petri dishes and covering them with light mineral oil to keep them separate and retard evaporation. Alternatively, a multiwell microtiter plate with sloping sides can be used, although the optics may not be optimal for scoring if an embryo settles near the edge, and care will be required to maintain adequate humidity to prevent evaporation. Equilibrate culture drops in a gassed incubator for an hour or so before placing embryos into the drops.

Handling embryos is best done with a mouth-controlled Pasteur pipette that has been pulled out over a flame and broken off at an internal diameter of about 120–150 μm (or just large enough that an embryo fits inside the pipette easily without distortion). Before picking up the embryos, fill the pipette up to the first widening of the pipette bore. You can pick up a large number of embryos at a time, but keep them near the tip so that they are not lost on the air/medium interface. Monitor all embryo manipulations under a dissecting microscope. Once the embryos are in their individual culture drops, they can be monitored and scored over a period of days. During monitoring, minimize time out of the incubator to avoid too many temperature and pH fluctuations.

membrane components. Feedback from the position of a cell with respect to other cells influences gene expression. The result of all of this is the reproducible differentiation of different cell types in appropriate places at appropriate times within the embryo. Certain features, such as compaction of the blastomeres and the pumping of fluid by the trophectoderm, occur at set times after fertilization and are independent of cell number. Other features, such as the differentiation of the ICM, depend on cell position and microenvironment and thus require a sufficient number of cells to produce the necessary microenvironment for the formation of the ICM. Once you locate the defective process of your mutant embryos, you will be able to refer to the associated literature to determine what is known about the biochemistry and genetic control of each process to get at the mechanism of action of your mutation. Some general methods for additional analysis of preimplantation embryos may be found in Section 5.C.5.

Some Possible Phenotypes

5.C.1.a No mutant embryos

By this stage of the game, you should have ruled out a haploid effect on sperm or oocyte, so a failure of fertilization should not be the problem. Some rare examples of sperm/oocyte incompatibility exist, but this is an unlikely possibility. Double check the accuracy of the genotyping procedure and collect more embryos to ensure that the assessment of no mutant embryos present is correct. If it is, perhaps you did not follow all of the diagnostic steps outlined in Chapter 3. If you determined that the male (or female) offspring of the chimera were fertile, but did not check the other sex as suggested in Section 3.D, it is just possible that you are only now turning up a dominant effect on oogenesis (or spermatogenesis). Go back to Section 3.D.

5.C.1.b Arrest or delay during cleavage

This will be evident as (1) a failure of the mutants to progress beyond a certain number of cells or (2) significant lagging of the mutants behind the wild-type and heterozygous embryos. Although it seems very easy to count cells (blastomeres) of precompaction embryos under stereo dissecting microscope optics, the morphology can sometimes be deceptive if the embryo is fragmenting, and once compaction starts (Figure 5.4F), it is nearly impossible to count cells this way. However, simple nuclear counting techniques with fluorescent dyes can be used that are much more accurate (Box 5.5; Figure 5.5). But remember that you may no longer be able to genotype an embryo that has been used for cell counting, so try to pin down the suspected phenotype first. The nuclear counting technique provides an accurate cell count and can also be used to recognize cell division and cell death by means of nuclear morphology so that you can assess proliferation and cell death.

An alternative to counting nuclei by using added nuclear dyes is to use certain nuclear-localized transgenic markers, such as histone-H2-eGFP (Figure 5.6), to accurately assess cell number, cell death, and mitosis, particularly when combined with confocal microscopy. The advantage of this type of marker is that the embryos can be observed vitally and recovered for further analysis or genotyping. However, the transgenic marker will have to be crossed onto the strain of mice carrying the mutant allele by breeding before it can be used.

Box 5.5 | Cell Counting Techniques for Preimplantation Embryos

Using fluorescent DNA dyes, simple nuclear counts can be done on preimplantation embryos for an accurate assessment of total cell number. Alternatively, a more complicated technique for differential counts of trophectoderm and ICM cell numbers can be done on blastocysts or even late compact morulae. The cell counting techniques have the advantage that you can distinguish interphase nuclei, mitotic cells, and dead cells by their characteristic chromatin patterns and thus quantitate mitosis and cell death. If you do not have a confocal microscope handy, however, you may have to squash the more advanced embryos to see all of the nuclei, in which case you will be unable to easily recover material for subsequent genotyping.

For simple total cell counts, expose the embryos to the DNA dye Hoechst 33258 in culture medium for 30 minutes and then examine them under UV light on a fluorescence microscope. A convenient way to mount the embryos that allows for the possible need to squash them if all of the nuclei cannot be easily distinguished is as follows:

- Have precleaned microscope slides, small square coverslips, and petroleum jelly handy. Rub a little petroleum jelly into the heel of your hand and scrape the edge of the coverslip gently across your hand to collect a thin ridge of jelly along one side; repeat on the opposite side and set the coverslips aside (jelly-side up).

- Pick up an embryo with a mouth pipette and place it on the microscope slide with about 10 µl of medium. Carefully place the coverslip over the embryo, making contact with the drop of medium but letting the coverslip rest on the petroleum jelly ridges. Observe the embryo under the fluorescence microscope and count nuclei.

- If you need to squash the embryo for better observation of nuclei, carefully press on the coverslip with a pair of forceps, keeping your eye on the embryo until it is sufficiently squashed. A few nuclei may shoot off if cells lyse, but you can scan the entire field under the coverslip for errant nuclei. If the drop is too big, it will be hard to squash the embryo sufficiently because of capillary forces once the fluid fills the space under the coverslip. If the drop is too small, the capillary force might overcome the resistance of the petroleum jelly and the embryo may be squashed with too much force. A brief pretreatment (20–30 seconds) with 1% Triton X100 will make the embryos easier to squash and prevent the distortion of nuclei-associated cell lysis.

For differential cell counts of trophectoderm and ICM, two DNA dyes are used: one that enters vital cells (Hoechst) and one that is excluded from vital cells (propidium iodide, PI). The technique uses a procedure called immunosurgery (MM3), which depends on the existence of tight junctions between cells of the trophectoderm. These tight junctions form during the compaction phase of preimplantation development, and thus differential cell counts can only be made on embryos that are compacted or have formed blastocysts. In this procedure, the embryos are first exposed to an antimouse antibody, e.g., rabbit antimouse RBC, rabbit antimouse embryo, etc. (you may have to beg, borrow, or make it yourself; we have not found a commercial source that works well). Then, wash and expose them to the two fluorochromes and a source of complement (guinea pig complement) that lyses the outer cells. The inner cells were protected from exposure to the antibody by the tight junctions in the trophectoderm and will therefore remain intact and exclude PI. The nuclei are detected as before, but this time the trophectoderm nuclei are pink (red plus blue) and the ICM nuclei are blue (Figure 5.5). Because the immunosurgically treated embryos are delicate and tend to fall apart, it is best not to put them through a wash step before counting nuclei. However, this means that the dyes will still be present in the medium, and, when you squash the embryos, the PI will rapidly enter the ICM nuclei because squashing damages the cell membranes. Count all of the trophectoderm nuclei before squashing and then be prepared to count ICM nuclei quickly, or take a photograph as soon as you squash before the ICM nuclei change color.

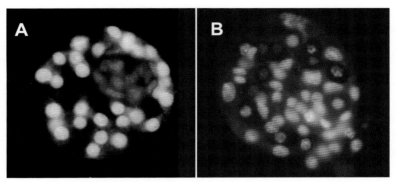

Figure 5.5. Differential cell counts of the ICM and trophectoderm of a double-dyed blastocyst before (*A*) and after (*B*) squashing. The trophectoderm nuclei appear to be reddish pink due to the PI and Hoechst dyes, whereas the vital ICM nuclei appear blue. (Reprinted, with permission, from Goldin and Papaioannou, *Genesis 36*: 40–47 [2003].)

5.C.1.c No compaction

Check cell numbers (Box 5.5) to determine if this is a case of developmental arrest before compaction, which should occur between the 8- to 32-cell stages, or whether cell number continues to increase in the absence of compaction. The compaction process relies on changes in cell adhesion, and if the blastomeres continue to divide but do not compact, alterations in cell adhesion could be the problem.

5.C.1.d Abnormal blastocyst morphology

If the embryos compact but do not form a cavity, it could signal a developmental arrest or a defect in the pumping mechanism of the trophectoderm cells. Trophoblast cells first form intracellular vesicles, which later coalesce to become the blastocyst cavity. Integrity of the blastocyst cavity depends on the tight junctions that form between adjacent trophectoderm cells. If a blastocyst cavity forms, but there is no ICM, this could be due to fewer-than-normal cells present at the time of compaction, in which case not enough cells are available to enclose cells in an inside microenvironment and all of the cells become trophectoderm. Alternatively,

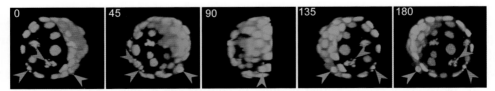

Figure 5.6. Nuclear-localized green fluorescent protein (GFP) as a cell counting tool. Three-dimensional reconstruction of half of a blastocyst transgenic for a histone-enhanced GFP fusion gene. Each panel shows a 45° rotation from the previous panel to show a 180° rotation of the half blastocyst. The inner cell mass is on the right of the first panel. With this chromatin-specific label, all nuclei can be counted in confocal images and, in addition, pycnotic cells (*orange arrowhead*), mitotic cells (*purple arrowhead*), and interphase nuclei can be distinguished. The polar body (*blue arrowhead*) is also visible. (Images courtesy of Kat Hadjantonakis and Virginia Papaioannou.)

the ICM might have formed but undergone cell death. The cell counting technique should sort this out. For further analysis, go to Section 5.C.5.

5.C.2 Techniques for the analysis of blastocysts and the process of implantation

If your mutant embryos form normal-looking blastocysts but fail to implant, several in vitro assays can pinpoint what is wrong. From the 3.5-dpc starting point (Figure 5.7A), culture blastocysts under conditions that will allow them to "hatch" from the ZP and attach to and outgrow on a culture dish (Box 5.4). Then, determine which aspect of blastocyst function is affected by the mutation.

Some Possible Phenotypes

5.C.2.a Failure to hatch

In the uterus, embryos eventually escape from the ZP, probably through a combination of blastocyst expansion and zonalytic activity in the uterus. In the artificial culture situation in vitro, escape from the ZP depends on the continued expansion of the embryo until it cracks the ZP and escapes through the crack. This "hatching" process is taken as a measure of a healthy, functional trophectoderm, and thus failure to hatch in vitro could reflect a trophectoderm defect, either through decreased cell number or decreased pumping activity. Unfortunately, the rate of spontaneous failure to hatch in vitro is fairly high, but if you can genotype your embryos, you can still correlate this phenotype with genotypes (e.g., perhaps 23% of wild-type embryos fail to hatch but 99% of homozygous mutants fail, indicating a clear, genotype-driven difference). If you cannot genotype the embryos, it is best to do a control backcross and compare the incidence of nonhatching with the rate among intercross embryos (see Box 5.3). Failure to hatch by itself is not a very satisfactory end point. If this is your phenotype, go on to Section 5.C.2.b. and eventually Section 5.C.3.

5.C.2.b Failure to attach

The next step is to determine if the mutant embryos attach to the culture dish. This process is mediated by the trophectoderm and is presumably the in vitro counterpart of trophectoderm attachment to the uterine epithelium, the first step in the implantation process. To give the embryos the best chance to show their capabilities, you may wish to eliminate the hatching hurdle if you have already established that your mutants hatch at a normal rate. In fact, even if your mutants show a hatching deficiency, it is worth removing the ZP to further test the functioning of the trophectoderm. To do this, simply remove the ZP (Box 5.6; Figure 5.7A,B) before you place the blastocysts in culture. Then sit back and wait for the mutants to appear. Culture the embryos individually, but identified by litter to confirm the mothers' genotypes. Score the embryos daily for attachment to the culture dish. If one quarter of the embryos continue to float while their sibs are attaching and outgrowing, you have probably found the phenotype. Genotype the embryos to confirm this and then go to Section 5.C.3. If you cannot distinguish the mutants, continue on to Section 5.C.2.c.

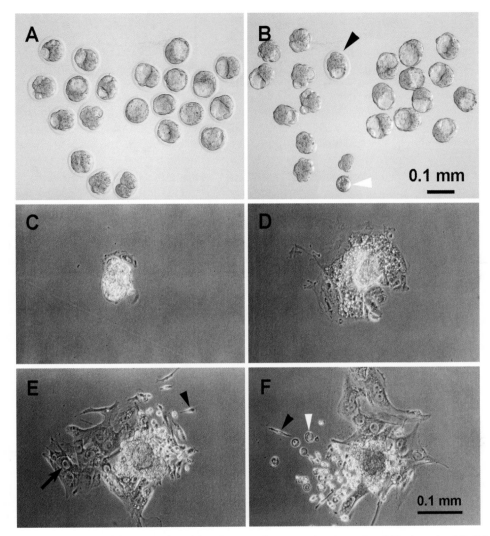

Figure 5.7. Zona pellucida removal and in vitro attachment and outgrowth of blastocysts. (*A*) 3.5-dpc embryos freshly flushed from the uterus. The embryos have been sorted by gross morphology: The 11 embryos on the right are expanded blastocysts; the seven on the left are beginning to cavitate or are partially expanded; the three on the bottom have not cavitated. (*B*) The same embryos after removal of the zona pellucida from all but one of the partially expanded blastocysts (*black arrowhead*). The embryo at bottom center has separated into a compact group of cells and a single arrested blastomere (*white arrowhead*). (*C,D*) Embryos explanted at 3.5 dpc after 2 days of in vitro growth. The blastocyst in *C* has collapsed and attached to the culture dish and the trophoblast cells are beginning to grow out from the central clump. In *D*, the trophoblast cells form a monolayer on the culture dish surrounding the central clump of cells composed of ICM derivatives. (*E,F*) Embryos explanted at 3.5 dpc after 7–8 days of in vitro growth. The large, clear nuclei with prominent nucleoli are the giant trophoblast cell nuclei (*arrow*) in the trophoblast monolayer surrounding the ICM. ICM derivatives form a compact central mass and have also begun to migrate away from the outgrowth (*black arrowheads*). In *F*, some rounded dead cells are evident (*white arrowhead*), a normal occurrence after this period of culture. Scale bar in *B* for *A,B*; scale bar in *F* for *C–F*.

> **Box 5.6** Removal of the Zona Pellucida
>
> The quickest and easiest method of removing the ZP, the glycoprotein coat surrounding the oocyte and early embryo, is to dissolve it in an acidic solution. The only trick is to do so without exposing the oocytes or embryos to the acid long enough to cause them damage. To accomplish this, use buffered embryo culture medium to control the microenvironment around the embryo. Perform the whole procedure under a dissecting microscope and your watchful eye in two watch glasses, one with buffered medium and the other with the acidic solution.
>
> - Use acidic Tyrode's salt solution (MM3) (pH 2.5) at room temperature. This solution is very stable and can be made in advance and either frozen as aliquots or stored at 4°C indefinitely.
> - Draw up a small amount of acidic Tyrode's solution into a mouth pipette, pick up the embryos a few at a time from a dish of buffered medium, and place them in the dish of acidic Tyrode's solution.
> - Keep a close watch on them and, at the same time, empty the mouth pipette and draw up a small volume of buffered medium.
> - As soon as you see the ZP start to disappear, expel a little of the buffered medium over the embryos and then collect them in the pipette and transfer them back to the dish of buffered medium. The ZP should be gone (see Figure 5.7A,B).
>
> If the ZP does not dissolve within 10–30 seconds in acidic Tyrode's solution, you may have transferred the embryos into the solution along with too large a volume of buffered medium. Move them around in the dish and next time use a smaller bore pipette or less medium. Do not be tempted to work with more than a few embryos at a time, because it can be hard to keep track of them in the dish. If you lose an embryo in the acidic solution, do not waste time searching for it—it will be beyond rescue.

5.C.2.c Failure to outgrow

The last phase of implantation in the uterus is the invasion of the maternal uterine epithelium by the trophectoderm and the concomitant transformation of the trophectoderm into trophoblast giant cells (cells that endoreduplicate their DNA in the absence of nuclear division, thus becoming very large in the process). The in vitro counterpart of this activity is the spreading of trophectoderm as a monolayer onto the culture dish and the loss of integrity of the blastocyst cavity. The ICM comes to lie on top of the trophectoderm monolayer in a small compound mound of cells, whereas the trophoblast giant cells, with their giant nuclei and refractile perinuclear granules, are easily visible in the surrounding monolayer. Monitor the explanted embryos daily and classify them into the following categories (Figure 5.7): expanded attached, attached but not expanded, outgrowing with ICM and trophectoderm giant cells visible, and outgrowing but with no ICM and/or trophectoderm. These classifications will help pinpoint a developmental delay, problems with the attachment or invasive phase of trophectoderm function, giant cell transformation, and development of the ICM. You could take closer note of the morphology, but outgrowing blastocysts are extremely variable, and, unless you are very experienced or have a control group growing in parallel, finer classifications that take into account the size or morphology of the ICM, etc., may not be worthwhile.

Starting at around 3.5 dpc and continuing through the implantation process, trophectoderm cells are highly phagocytic, and this forms the basis for an in vitro test of trophectoderm differentiation and function. Simply add 1- to 3-μm fluorescent or colored-latex microspheres to blastocyst cultures for 8 hours or more, wash the cultures, and determine whether the trophectoderm cells have phagocytosed any of the spheres. This can be applied to zona-free blastocysts or to embryos that have attached to the culture dish.

If you have reached this section with a diagnosis of a preimplantation lethality, we fully expect that you will have found your phenotype by this point, because it is unlikely that the homozygous mutant embryos even got as far as attaching to the culture dish. We say this because attachment usually stimulates the uterus to undergo a semiautonomous process called decidualization. Once the uterus has received the stimulus of an attaching embryo, the process does not depend on the continued presence or normality of an embryo; in fact, an oil drop, mechanical injury, and even certain bacteria can stimulate a decidual reaction. Thus, even a defective embryo could induce decidualization as long as it can hatch and the trophectoderm can start to attach. It is the decidua that you were looking for as a sign of the beginning of implantation in Section 5.B and you have reached this section because an insufficient number of empty decidua were present to account for your mutant embryos. Therefore, it stands to reason that the mutants will probably not attach or outgrow in this in vitro assay. However, you can still do several more tests if blastocysts failed to attach, so go on to Sections 5.C.3–5.C.5.

5.C.3 Analysis of isolated ICM

Because there are two distinct tissues present in the blastocyst, it is possible that the gene you are investigating has its role in only one or the other cell type. If you have expression data, you might even have a good idea of which cell type is most likely to be affected. The assays in Section 5.C.2 mostly test the trophectoderm functions of attachment, invasion, and differentiation. But even if you have already identified a trophectoderm defect that resulted in failure of implantation, there may also be independent effects on the ICM. So now let us see what the ICM can do on its own. First remove the ZP (see Box 5.6). Then, the same immunosurgical technique that was described in Box 5.5 (but without the fluorescent dyes) can be used to isolate the ICM from the trophectoderm (Figure 5.8). In this case, the expanded blastocyst is exposed to antimouse antibody, washed, and then exposed to complement, which will lyse all the trophectoderm cells. Gentle pipetting with a finely drawn, mouth-controlled pipette will remove the trophectoderm debris, leaving a pure ICM that you can culture in vitro (see Box 5.4).

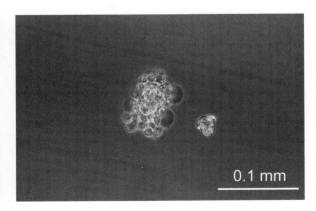

Figure 5.8. Isolation of the ICM from 3.5-dpc blastocysts. Blastocysts were treated with rabbit antimouse antiserum, washed, and exposed to guinea pig complement. The outer trophectoderm cells in the embryo on the left are undergoing lysis. The isolated ICM on the right was separated from the layer of dead trophectoderm by gently pipetting.

Around the time of implantation, the ICM forms a layer of primitive endoderm on its blastocoelic surface in the second cellular differentiation step of embryogenesis. This also occurs in isolated ICMs in culture, so that after about 2–3 days (3.5 dpc plus 2–3 days of culture), the endoderm forms a rind around the spherical ICM. A basement membrane will be evident at the basal surface of the epithelial endoderm cells. A possible phenotype to look for is the absence of a primitive endoderm layer, which should be detectable either by morphology or by primitive endoderm marker gene expression. Its absence could indicate a developmental arrest or a failure of primitive endoderm differentiation. Another possibility is that the mutant ICMs fail to thrive in culture and either die or arrest cell division. The cell counting technique for preimplantation embryos (see Box 5.5) can be applied to isolated ICMs as well. Keep in mind that in the intact embryo, failure of the ICM will have secondary effects on the trophectoderm because a signal from the ICM is required to keep the trophectoderm in a proliferative state to form the ectoplacental cone and eventually the placenta. In the absence of contact with a functional ICM, the trophectoderm forms only nondividing, terminally differentated giant cells, which are able to elicit a decidual response but will not support further development.

5.C.4 Analysis of implantation delay

Many mammals, including mice, have a reproductive adaptation known as implantation delay, or diapause. Basically, preimplantation embryos pause in their developmental program, often due to some adversity, and await a more propitious time to resume development. This is actually a state imposed by the reproductive tract of the mother when nutritional or other conditions do not favor the embryo. In the case of mice, implantation delay can occur while the mother is nursing a previous litter. Other mammals such as some marsupials, bears, and seals make implantation delay a regular part of their reproductive cycles to coordinate reproduction with seasonal and environmental changes.

Because implantation delay is within the normal repertoire of behaviors for a mouse embryo, one can exploit this feature to help uncover cryptic phenotypes caused by mutant genes. Diapause can be induced experimentally by ovariectomy of the mother during the preimplantation phase of development. The lack of ovarian hormones causes the embryos to enter the physiological state of diapause, during which they are somewhat quiescent and do not initiate implantation. They do, however, hatch from the ZP and continue to increase their cell number, albeit slowly. Maintenance of viability and developmental potential during diapause has a genetic basis (like everything else) and mutation in at least one gene (*Lif*) reveals a role in maintenance of the ICM during diapause when all other aspects of the mutant embryo appear normal. Thus, you may wish to try implantation delay in your investigation under two conditions: (1) if your gene is expressed during preimplantation or periimplantation development but you do not see a phenotype at that time or (2) as an additional test of function and viability in embryos that do have a blastocyst or periimplantation phenotype.

Ovariectomize pregnant females (MM3) and count CL (see Box 3.3, p. 33) so you know the maximum number of embryos that you can expect to recover later. This procedure will cause the embryos to arrest in diapause or implantation delay at the blastocyst stage, although they will shed the ZP and increase cell number slowly. After 4–8 days, recover embryos by flushing the uterus as you would for normal blastocysts. Delayed embryos can survive longer in the uterus,

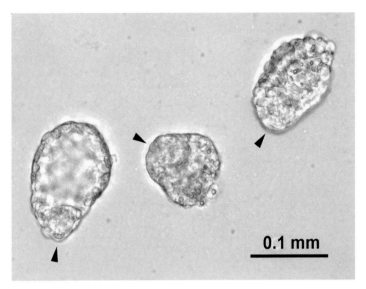

Figure 5.9. Blastocysts in implantation delay. The three blastocysts are in diapause or implantation delay after recovery from the uterus at 6.5 dpc following ovariectomy at 2.5 dpc. The embryos have escaped from their zonae pellucidae and giant cell transformation has begun at the abembryonic pole in the two embryos on the right. Arrowheads indicate the embryonic pole with the inner cell mass.

but they will begin to decline after about the first week so the analysis should be done sooner rather than later. The morphology of delayed embryos can be quite variable. They might not be expanded, and frequently, giant cells will be evident in the mural trophectoderm (Figure 5.9). Assessment of these embryos for possible mutant effects could include determining embryo loss (by comparing numbers recovered with the CL counts), simple or differential cell counts (see Box 5.5), and assessment of cell death; determining further developmental potential in vitro (see Box 5.4); and assessing differentiation with molecular markers. Now go on to Section 5.C.5 below.

5.C.5 Further analysis of preimplantation embryos

The previous sections rely heavily on morphological assessment of living embryos isolated at different preimplantation stages or cultured in vitro. Combined with staging, cell counting, and measurement of cell death, this analysis should provide a clear assessment of the nature of a preimplantation or implantation defect. Further analysis will depend on what you have learned and the nature of the gene mutated. The next level of analysis combines morphological analysis at the histological or ultrastructural level with various means of determining the localization of mRNA and proteins, as well as the more global assessment of gene expression patterns. Whenever possible, ascertain the genotype of each embryo, and, if this is not possible, compare with a control group of embryos from heterozygous × wild-type mice (see Box 5.3) and test a large enough group of embryos to attain significance in a χ^2 analysis (see Box 5.1). Superovulation of females (MM3) is a useful means of increasing embryo number for preimplantation analysis. If there is any reason to suspect that the sex of the embryo affects the phenotype or the genotype affects the sex, the embryos can be genotyped (Box 5.7).

Box 5.7 Sexing Mouse Embryos

Knowing the sex genotype and phenotype of embryos can be essential if your mutant gene is suspected of affecting sexual development or could in any way be sex limited. An XX genotype should be female (i.e., have ovaries) and an XY genotype should be male (i.e., have testes). Mice can have sex karyotype variants, including XO females and XXY males. There are also many instances of intersex phenotypes including true hermaphroditism (mice with ovotestes) and pseudohermaphroditism (mice with either testes or ovaries but both male and female reproductive organs). In most instances, a complete sex reversal (XX male or XY female) or an intersex phenotype will be infertile. Many methods exist for determining the sex of embryos: some that are appropriate for any age and some that are age specific.

Genotype (XX versus XY)

- PCR primers for genes located on the Y chromosome can be used on any embryonic tissue at any age (e.g., *Sry* and *Zfy*) (see Box 3.7, p. 40).
- The Y-chromosome-specific repetitive element Y353/B probe can be used on Southern blots or dot hybridizations on any embryonic tissue at any age (MM3).
- The Y-chromosome and chromosome-19 autosomes are the smallest mouse chromosomes. Thus, a male karyotype will have three very small chromosomes and a female karyotype only two (MM3).
- Y-chromosome-specific probes can be used on chromosomal spreads using fluorescence in situ hybridization (FISH) from any age embryo.
- Between 9.5 and about 12.5 dpc, sex chromatin can be detected in the amnion of female embryos (Figure 5.10A,B), leaving the rest of the embryo intact for additional analysis.
- XX and XY embryos can be distinguished after the blastocyst stage in crosses using males carrying an X-linked *eGFP* transgene crossed to wild-type females. All of the daughters from this cross will be eGFP positive ($X^{eGFP}X$) and all of the sons eGFP negative. This method of genotyping is useful because you can easily and quickly sort living embryos under a fluorescent dissecting scope to pool the different sexes for further analysis.

Phenotype (female versus male)

From about 12.5 dpc, the seminiferous cords become apparent as the testes differentiate. At this stage, the fetal ovaries appear homogeneous in structure. The exact timing of the differentiation of the testis can be somewhat variable, depending on genetic background and the light schedule of your mouse colony. From 13.5 dpc onward, the morphological differences between testes and ovaries are clear (Figure 5.10C,D). At late fetal stages as well as at birth and at all subsequent stages, morphological features of the external genitalia distinguish male and female mice (see Figure 3.5).

5.C.5.a Histology of preimplantation embryos for histochemistry and in situ localization of mRNA or proteins

Cleavage-stage embryos can be fixed and processed within oviducts removed intact from pregnant females. Once embryos have moved through the uterotubal junction and dispersed throughout the uterus, however, it becomes impractical to section the entire uterus to find them. A good alternative is to collect blastocysts by flushing them from the uterus as usual and then transferring them into an isolated oviduct (you can use one from the same female) as if you were doing embryo transfer (MM3). The oviduct acts as a container for fixing and processing

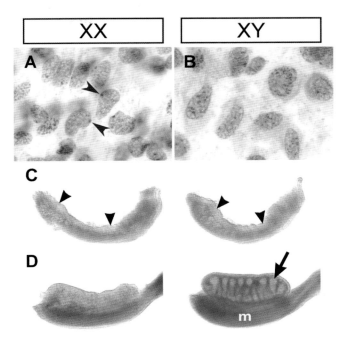

Figure 5.10. Sexing mouse embryos. (*A,B*) Sex chromatin in the amnion of 10-dpc embryos. Isolated amnions (MM3) are spread on a glass slide and stained with a drop of 2% orcein in 60% acetic acid. Nuclei of the amnion cells of the female (*A*, XX) have a spot of darkly staining chromatin on the nuclear membrane (*blue arrowheads*), which represents the inactive X chromosome. In humans, this chromatin is known as the Barr body. No such spots are seen in the male nuclei (*B*, XY). (*C,D*) Dissected urogenital ridges from 11.5-dpc (*C*) and 13.5-dpc (*D*) female (XX) and male (XY) embryos. The gonads, which are located on the medial surface of the mesonephroi, have a transparent appearance. (*C*) At 11.5 dpc, the male and female gonads are morphologically identical. (Arrowheads indicate the rostral-caudal extent of the gonads.) (*D*) 13.5-dpc female and male gonads with attached mesonephroi (m). At this stage, the male and female gonads are morphologically distinct. The testes have cords (*arrow*), whereas the ovaries do not. (*C,D* provided by Akio Kobayashi and Richard Behringer.)

a large number of embryos in a compact package that is easily handled and sectioned. Either paraffin or plastic embedding can be used, depending on the intended use and level of detail required, the latter providing higher resolution of cellular detail. Sections can then be subjected to specific histochemical staining or immunocytochemistry to highlight cellular components and obtain a high-resolution morphological picture, perhaps by taking advantage of the use of confocal microscopy. The expression pattern of specific genes and the localization of specific gene products can be determined using in situ hybridization or immunolocalization. A number of molecular markers of different cell types or different states of differentiation will aid in characterizing the mutant embryos, depending on the phenotype.

HELPFUL HINT

When using reverse transcriptase–polymerase chain reaction (RT-PCR) for genotyping preimplantation embryos, it may be necessary to remove the zona pellucida to avoid possible contamination from cytoplasmic remnants of cumulus cells trapped within the ZP.

5.C.5.b Gene expression in whole embryos

Gene expression analysis or immunolocalization of proteins can also be done on whole preimplantation embryos, although the embryos can be tricky to handle through all steps of the procedures. In addition, the resolution will be less than that obtained from sections. On the other hand, embryos processed as whole mounts are available for genotyping afterwards, making the number requirement much lower for a definitive assessment of phenotype. Alternative means of assessing mRNA or proteins in individual embryos include RT-PCR to detect specific transcripts of interest, or two-dimensional gel electrophoresis of radiolabeled proteins. Because the analysis must be done on a single-embryo basis, the sensitivity of the different assays can be limiting. Every attempt should be made to develop a means of genotyping embryos at the same time to attribute differences definitively to the homozygous mutant phenotype.

5.D Analyzing Periimplantation and Early Postimplantation Phenotypes Lethal between 4.5 and 9.5 dpc

Indications for embryonic lethality during early gestation are that (1) no normal, living homozygous mutant embryos are recovered from dissections at 12.5 dpc and (2) the number of implantation sites that are either empty of embryonic remains or have only fetal membranes and degenerated fetuses fits the expected number of homozygous mutants. If you were successful in genotyping any of these abnormal remains, they should be the homozygous mutants, although there is likely to be some low level of embryonic wastage due to causes unrelated to the mutation. The next step is to pin down the time of onset of the phenotypic effects by further dissections and embryo collections, working backwards to find the time at which all embryos, including homozygous mutants, are normal in size and morphology (Figure 5.11). You may wish to sex the embryos to ensure that you have a normal sex ratio among mutants (see Box 5.7). Even if you do not reach the point of seeing normal embryos, estimating the approximate time of death from the nature of the embryonic remains (see Section 5.B, Step 4) will indicate how you should proceed: With no embryonic remains or only a few giant trophoblast cells and membranes in a small decidual swelling, periimplantation lethality is indicated. Go to Section 5.D.1. For later times of death indicated by the presence of recognizable dead fetuses or degenerating fetal remains, go to Section 5.D.2.

5.D.1 Periimplantation death, 4.5–5.5 dpc

If small decidua with no embryonic remains or only a few giant trophoblast cells and some membranes are found at 12.5 dpc, the indication is for death shortly after implantation. The most common cause of lethality at this stage is failure of the embryo to implant in the uterus. Because trophectoderm growth and continuing function depend on the presence of a viable ICM, lethality at this stage could be due to a mutant effect in either trophectoderm or ICM. To investigate this phenotype, gross dissections will be impractical for all but the experts. You can, however, flush 4.5-dpc embryos from the uterus in the same way that earlier-stage blastocysts are flushed. Depending on the developmental stage, some or all of the embryos may be in the

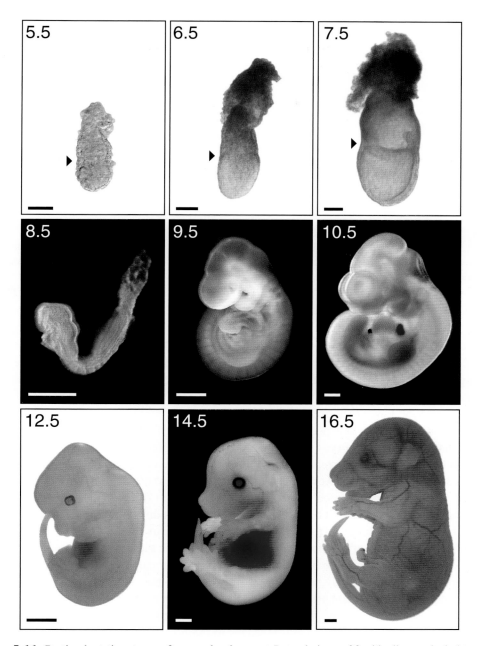

Figure 5.11. Postimplantation stages of mouse development. Lateral views of freshly dissected whole-mount embryos at various days postcoitus (dpc), indicated in upper left corner of each panel. All embryos are oriented with anterior to the left. Arrowheads in 5.5- to 7.5-dpc images indicate junctions between the embryonic (*below*) and extraembryonic (*above*) regions. The trophoblast and Reichert's membrane have been removed from the 5.5- to 7.5-dpc embryos. The placenta, Reichert's membrane, yolk sac, and amnion have been removed from all of the other embryos. Bars, 0.1 mm, 5.5–7.5 dpc; 0.5 mm, 8.5–10.5; 1 mm, 12.5–16.5. (Image of 5.5-dpc embryo provided by Aya Wada and Richard Behringer.)

96 | Chapter 5

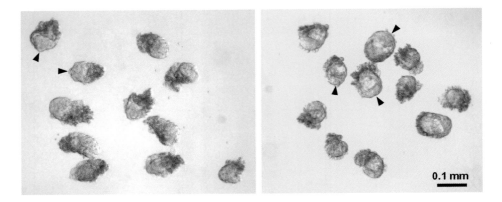

Figure 5.12. Blastocysts flushed from the uterus at 4.5 dpc. (*Left*) Implanting embryos that were forcibly flushed from the uterus. (*Right*) The same embryos after one hour of culture. At initial recovery, the embryos are collapsed, but many pump up to a more recognizable blastocyst morphology following culture. The giant cells are evident at the abembryonic pole, opposite the smooth embryonic pole containing the ICM (marked with an arrowhead in several embryos). Some maternal uterine epithelial cells are still attached to the trophectoderm.

process of attaching to the uterine wall, so a more forceful flow of medium through the uterus may be necessary to dislodge them. Typically, the uterus will balloon with the pressure of the flushing medium and then the fluid will squirt out as the implanting embryos are dislodged. The rate of recovery of embryos using this method is not as high as that for preimplantation stages and may be lower than 50%, but CL counts can still provide an estimate of the maximum number of embryos to expect, thus providing a good estimate of the recovery rate. Keep in mind that the mutant embryos may be lagging in development and may not have begun to attach, and thus could be overrepresented among the recovered embryos. Accordingly, genotyping is as important as ever. At this stage, wild-type embryos are no longer ZP contained and will have giant trophectoderm cells at the abembryonic pole (Figure 5.12). They can be morphologically quite variable, which makes picking out a defect difficult unless it is striking (e.g., no ICM or severe retardation). Differential cell counts (see Box 5.5) might provide a clue, but the technical difficulties of counting cells and genotyping the same embryo means that a large number of embryos will need to be counted and possibly compared with a control population.

If everything looks relatively normal at 4.5 dpc, but you know that the embryos die at periimplantation, either the mutant effect appears later or it is not evident by gross morphology (or cell number). Rather than trying to dissect 5.5-dpc embryos, skip directly to histological evaluation of the embryos at periimplantation stages in utero (see Box 5.8) and carry out a functional assessment of the preimplantation and periimplantation embryos to uncover hidden defects and determine the time of onset of the mutant effect. Start with an assessment of blastocysts at 3.5 dpc (Section 5.C.1.d). If they appear normal, continue by testing in vitro outgrowth (Section 5.C.2.c), ICM development (Section 5.C.3), and possibly implantation delay (Section 5.C.4). If, on the other hand, the blastocysts already appear abnormal by 3.5 dpc, they have preimplantation problems even though they are capable of eliciting a decidual response. In this case, go back to Section 5.C.1.

5.D.2 Gastrulation to allantoic fusion, 6.5–9.5 dpc

If embryonic lethality after the periimplantation period is indicated by the presence of degenerating embryos or extraembryonic tissue at 12.5 dpc, select alternate days (e.g., 7.5 and 9.5 dpc) for the next dissections. Dissect and examine each implantation site as before (Section 5.B), using techniques appropriate for each stage (MM3), and genotype from whatever embryonic or extraembyronic tissue is available, except trophoblast, which is likely to be contaminated with maternal tissue. When you find a stage at which all homozygous mutants are normal in size and morphology, you have identified the starting point for further analysis and

---

Box 5.8 Orientation of Embryos for Histological Sectioning: Embryos in the Uterus

Because embryos have a predictable orientation with respect to the uterus (MM3) (Figure 5.1), the key to preparing serial sections of embryos between 4.5 dpc and late gestation is to leave them in the uterus and use uterine landmarks to orient the tissue in the paraffin block during embedding. An added advantage is that the placenta and extraembryonic membranes will also be intact and in a suitable orientation for comparison and analysis.

- Dissect out the uterine horns along with a bit of the cervix and the oviducts, leaving some of the mesentery for orientation.

- Spread the uterus onto a small piece of dry index card and, with dissecting pins, pin it in place through the cervix and oviducts (Figure 5.13A). The tendency of the uterine muscle is to contract, so stretch each uterine horn out straight (but not under too much tension) and spread the mesentery out on the card to anchor the uterus on the paper. This will result in all of the embryos being in more or less the same orientation, with the placenta on the mesometrial side of the uterus.

- Slip the uterus, card and all, into fixative (Bouin's [MM3] is our favorite for general hematoxylin & eosin [H&E] histology, but other fixatives such as 4% paraformaldehyde [PFA] may be required for other procedures such as in situ hybridization or immunohistochemistry).

- After fixation and before clearing and embedding (MM3), the uterine horns can be cut with a scalpel into convenient lengths for sectioning. At 4.5 dpc, the decidual swellings may not be visible, so each horn can be cut into segments about 1.5 cm in length; at later stages, 2–5 decidual swellings can be included in each segment, depending on their size, thus including 2–5 embryos in each block. This greatly decreases the amount of sectioning that needs to be done.

- The uterine segments can be positioned in any orientation in the paraffin blocks, but we recommend sectioning parallel to the long axis of the uterus, because sectioning perpendicular to the long axis will require many more sections to be cut. Position the uterine segments in the paraffin block as shown in Figure 5.13B, using the mesentery for orientation, with a label on the left-hand side. When sectioning, mount the block with the label on the left and make a diagonal cut in the top right-hand corner of the block (Figure 5.13C) to indicate the orientation of the ribbon after cutting. Mounting the ribbon on the slide as shown in Figure 5.13D results in sections of two (or more) embryos in rows on the same slide.

- The long axis of the mouse embryo, before turning, is more or less at right angles to the long axis of the uterus, with the dorsal side of the embryo toward the mesometrial pole. Thus, sections oriented as shown will be transverse or sometimes sagittal. Once embryos have turned, their orientation will be more variable, but the placenta is always at the mesometrial pole and will be nicely sectioned for analysis.

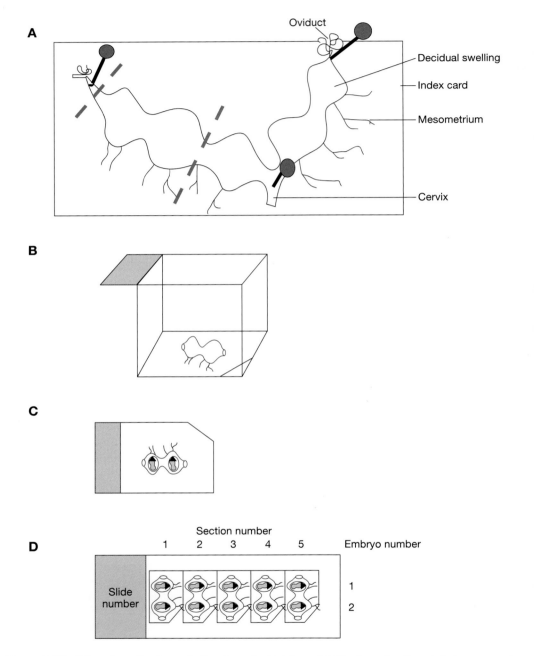

Figure 5.13. Orientation of embryos in the uterus for histology. (*A*) Pin down the dissected uterus on a piece of index card with dissecting pins, spreading the mesentery on the paper. After fixation, cut the uterus into conveniently sized segments (*dashed lines*); two decidual swellings are shown here. (*B*) Label embedding molds on the left-hand side and position the uterine segments with the mesentery at the front. (*C*) For sectioning, mount the block with the label on the left and trim the block, putting a notch in the top right-hand corner for orientation. (*D*) Mount the ribbon on labeled slides as shown. The first section is on the left and each of the embryos is in a horizontal row with the mesometrial pole, or placenta, on the right.

can move later in development until you encounter the onset of the abnormality. During the time frame of 6.5–9.5 dpc, some of the common causes of lethality are failure of gastrulation, failure of extraembryonic membranes to function correctly (including failure of vasculogenesis or hematopoiesis), failure of the allantois to fuse with the chorion, and lack of proper cardiac specification or cardiovascular failure, so be on the lookout particularly for these defects. The following techniques are some of the most common and useful means of analyzing these early-gestation stages. As you progress in the analysis of a phenotype, you may find that you wish to examine an earlier or later stage, so use these methods selectively but freely to investigate any features, or follow any intuitions you might have.

5.D.2.a Morphology: Gross and histological

The dissections done to pinpoint the time of death provide a gross morphological assessment of the phenotype over the time course of the embryonic lethality. This phenotypic and genotypic classification results in numbers that can be used to determine whether the onset of mutant effect occurs at a specific stage or is variable. The value of a good gross morphological assessment of freshly dissected living embryos should not be underestimated. The better versed you are in embryo development, the better prepared you will be to pick up deviations. Careful observation of embryos during dissection can reveal features that are not evident once they are fixed, including, for example, the presence, strength, and regularity of a heart beat; blood circulation and the vascularity of the placenta and yolk sac as seen by their color and the presence of blood in vessels; the presence of extravascular blood, which could indicate hemorrhage but might be washed away during tissue processing; or the condition of advanced necrotic embryos that cannot be further processed.

Sometimes a fragment of tissue will be recognizable in a necrotic embryo and can provide a developmental landmark. For example, eye pigment persists in degenerating embryos and indicates that the embryo reached at least 10.5 dpc before its death. Take detailed notes during dissections and photograph the embryos to document any morphological defects (Box 5.9).

The developmental stage of each embryo should also be determined (Boxes 5.10 and 5.11; see also Box 5.2). At the early end of this time range (6.5–7.5 dpc), the shape and size of the embryo can be used for staging as well as the extent of the primitive streak, the presence of the node, and the presence and size of the allantois (Figure 5.14). Toward the later end of the range (8.0–9.5), somite number is an easily assessed and accurate indicator of developmental stage (Box 5.11).

A range of developmental stages is commonly observed between and even within litters, but a large disparity could indicate that the mutation is causing a developmental delay. To the extent possible, you may also wish to determine that the rate of development of different organs within embryos is appropriate, particularly if the expression of your gene is localized there. For example, if the somite number is appropriate for a given age, is eye development also appropriate for that age or do the eyes lag behind? Be on the lookout for size differences within a given developmental stage. A mutation might cause a reduction in overall cell number, resulting in an embryo of small size but otherwise normal morphology. Growth retardation can also be the result of cardiovascular insufficiency.

Examine the heart to determine whether the bilateral heart tubes have migrated to the midline and fused to form a single heart tube by 8.5–9.0 dpc (see Figure 5.23, below). Irregular beats begin in the endothelial strands at about 8.0 dpc and the heartbeat should be strong and

Box 5.9 Photographic Documentation of Embryos

Good photographs enhance publications and are useful for documentation and analysis. When you are dissecting fresh embryos, they are usually destined for additional analyses and will never look the same again. It is not necessary to photograph every embryo, but it is worthwhile to take a picture of a representative wild-type phenotype as well as any abnormalities you detect. Even if you do not know what you are looking for yet, you can always refer back to photographs that you have taken. Documentation of in situ hybridization and other types of analysis is also worthwhile (MM3), even if the preparations are more or less permanent. Take the best-quality pictures you can each time (so that you will have publication-quality photos when the time comes) and maintain a detailed and rigorous bookkeeping system to keep track of them. The most convenient way to do this is with a computer with a high-quality digital camera that can be attached to both a dissecting and a compound microscope. As you photograph, store each image with a unique file name and record the details on a spreadsheet.

For fresh preimplantation embryos, phase or DIC optics are both good. Focus on the ZP until you see a sharp edge, which will be the equatorial plane. This provides the most pleasing image. Older embryos are best photographed on a dissecting microscope for which the depth of field is quite large (lower magnification will provide greater depth of field so that more of the three-dimensional embryo appears in focus). Embryos should be completely immersed in clean medium or phosphate-buffered saline (PBS) to avoid reflections from the surface. Experiment with backgrounds and lighting to get the best out of your particular sample and microscope. A useful setup for whole embryos is a transmitted light base with a tilted mirror and frosted glass stage, with additional light provided by a fiber optic illuminator with adjustable "gooseneck" arms. Adjust both the mirror to provide uniform background lighting and the fiber optic lights to provide appropriate top or side lighting. Then, adjust the light levels until the image shows the features that you wish to illustrate: Internal features will benefit from more transmitted light; surface features will benefit from more top lighting.

A neat trick that results in a nice uniform background and allows for flexible positioning of embryos is the use of a petri dish containing semisolid agarose. Prepare 60-mm plastic culture dishes with a layer of 0.5–1% agarose ahead of time and store them at 4°C. When you are ready to photograph embryos, simply flood an agarose dish with PBS or medium and introduce the embryos. To position the embryos, scoop out a shallow hole in the agarose and wedge the embryo into it in the desired position. This is especially useful for photographing at odd angles or for producing a lineup of embryos all in the same position, because it prevents one embryo from floating off as you position another.

Figure 5.14. (*Legend continued on following page.*) Downs and Davies staging system for 6.5- to 8-dpc embryos. All embryos are shown in side view with anterior to the left and ectoplacental cone removed. Preprimitive-streak (PS) embryos consist of two layers of cells with a circumferential constriction between the embryonic (e) and extraembryonic (x) regions, but no primitive streak. Early-streak embryos (ES) have a primitive streak at the posterior pole and a small wedge of mesoderm (m, *gray*) at the posterior pole. This is evident as an area in which the two layers, primitive endoderm and epiblast, are no longer distinct. Midstreak (MS) embryos are characterized by a primitive streak that has extended 50–100% of the length of the embryonic portion at the posterior pole. Late-streak (LS) embryos have a node (n) at the anterior end of the primitive streak, and also a posterior amniotic fold (paf) at the posterior end of the primitive streak, just at the boundary of the embryonic and extraembryonic regions. Neural-plate-stage embryos are staged on the basis of the allantoic bud: No bud (0B) embryos have no allantois but are characterized by the fusion of the amnion-

Primitive-streak stages

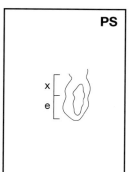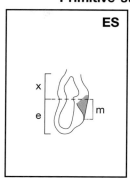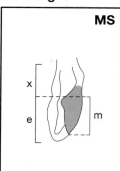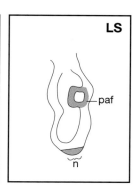

Neural plate stages

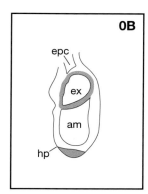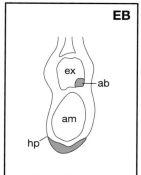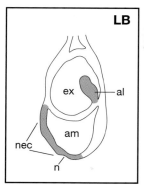

Headfold stages

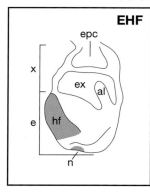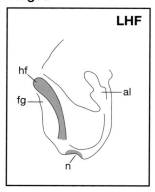

ic folds to form three separate cavities: the amniotic cavity (am), the extraembryonic coelom (ex), and the ectoplacental cavity (epc), and an anterior extension of the node to form the head process (hp). Early allantoic bud (EB) embryos have a small allantoic bud (ab) pushing into the extraembryonic coelom that is best seen in a posterior view (not shown). Late bud (LB) embryos have a long allantois (al) extending through the extraembryonic coelom and a thickened neural ectoderm (nec). The headfold (hf) stage can be subdivided into early headfold (EHF), which has bilaterally thickened headfolds anterior to the crescent-shaped node, and late headfold (LHF), which is characterized by larger, sigmoid-shaped headfolds. (fg) Foregut. (Adapted, with permission, from Down and Davies, *Development 118:* 1255–1266 [1993 ©Company of Biologists].)

regular by 9.5 dpc. Heart looping takes place starting at around 8.5 dpc, although defects in looping per se are not life threatening.

> **HELPFUL HINT**
>
> *Dissection in cold PBS can result in a temporary slowing or cessation of the heart beat, which is restored as the embryos warm to room temperature. The normal rate is 100–110 beats/minute. Note that because of the contractile nature of heart tissue, the presence of a heart beat is not by itself proof of a healthy embryo.*

Box 5.10   Downs and Davies Staging System for Preprimitive-streak- to Headfold-stage Embryos (6.5–8.0 dpc)

Between the preprimitive-streak stage, when the embryo consists of two cell layers in a simple cylindrical shape, and the headfold stage, when the embryo has three germ layers and all of the fetal membranes, growth and morphogenesis are rapid and embryos within a litter or age range can vary considerably. Because so many new tissues are appearing, it may be important to stage the embryos before making comparisons between mutant and wild type to detect a phenotype. Downs and Davies have described a staging system that divides this period into nine stages based on major morphological characteristics that are easily identified in intact embryos under a dissecting microscope, once Reichert's membrane has been removed. The primitive-streak stage is divided into four substages based on the presence and extent of the primitive streak; the neural-plate stage is subdivided into three substages based on the presence and growth of the allantois; the headfold stage is divided into early and late stages depending on the morphological appearance of the headfolds (Figure 5.14).

Box 5.11   Counting Somites as a Measure of Developmental Stage

Somites are the transitory, paired, segmental blocks of paraxial mesoderm that differentiate into the dermatome, sclerotome, and myotome, and eventually give rise to the dermis, vertebral column, ribs, associated musculature, and other structures. They form and then differentiate in an anterior-to-posterior progression. Between the ages of 7.5 and 11.5 dpc, their number provides a useful and fine-grained indicator of developmental stage and is especially useful when combined with measurement of crown–rump length to access body size. The first ten somites form at about one per hour, the trunk somites take between 1.5 and 2 hours each, and the tail somites form at a rate of about one every 2–3 hours. Counting somites in embryos of up to about ten somites is straightforward and the first somite is located just caudal to the otic placode (later, the otic vesicle). At later stages, the differentiation of the most anterior somites makes them difficult to discern; however, the position of somites relative to the emerging limb buds is constant, so the numbers can be accurately estimated. Somites 7–12 are in line with the forelimb bud, and somites 23–28 are in line with the hindlimb bud (Figure 5.15). Thus, in later embryos, you can start counting at one of these landmarks. These relative positions shift later in development but the method is valid for embryos up to about 40–45 somites (11.5 dpc). If in doubt, you can always count somites after using an in situ marker of somites such as *Meox1 (Mox1)*.

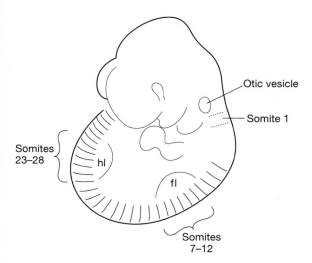

Figure 5.15. Counting somites to stage embryos up to 11.5 dpc. The first somite (*dotted lines*) is located just posterior to the otic vesicle but has differentiated by the stage shown and is no longer visible. Somites 7–12 are located at the level of the forelimb (fl) bud and somites 23–28 are located at the level of the hindlimb (hl) bud.

Examine embryos for laterality defects, which is best done at 9.5 dpc or later, after axial rotation from dorsal to ventral flexure. The most obvious asymmetry is cardiac looping but several laterality characteristics can be scored (Figure 5.16 and also see Figure 5.23, below):

- the heart makes a C-shaped loop to the right side;
- the tail and the placenta are normally on the right side of the embryo;
- the vitelline vessels emerge from the left side of the embryo and are always on the opposite side from the placenta;
- the spleen and stomach are located in the left side of the abdomen.

Finally, determine whether neural tube closure is taking place on schedule. Closure of the neural tube is a continuous process starting at the 6- to 7-somite stage (~8.5 dpc) at the level of the hindbrain (closure 1). Two other points of closure appear at the 12- to 15-somite stage

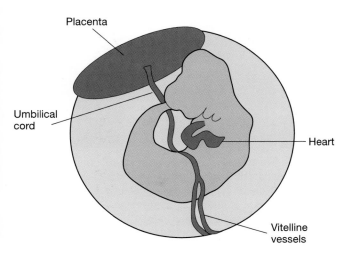

Figure 5.16. Laterality features of a 9.5-dpc embryo. The placenta and tail are on the embryo's right and the vitelline vessels emerge on the left. The heart has a C-shaped loop to the right. The outflow tract of the heart is on the right at the anterior pole of the heart tube. The inflow tract is at the posterior pole of the heart on the embryo's left.

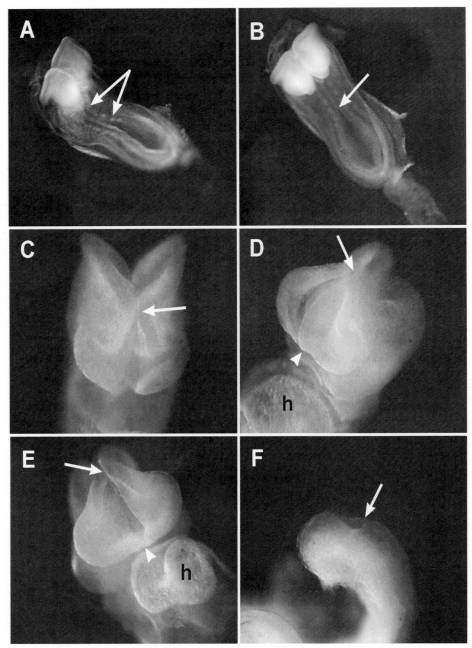

Figure 5.17. Neural tube closure. (*A,B*) Oblique and dorsal images of a 6- to 8-somite-pair embryo showing the first neural tube closure at the level of the future hindbrain–cervical junction (*arrows*). (*C–E*) Three different views of the head of a 12- to 16-somite-pair embryo showing the second neural tube closure at the level of the forebrain–midbrain junction (*arrow*) and a third point of closure at the rostral end of the neural tube (*arrowhead*). (*F*) The posterior neuropore at the caudal extremity of the neural tube. Closure of the spinal neural tube proceeds from the hindbrain–cervical junction caudally and eventually closes the posterior neuropore at about the 24- to 26-somite-pair stage. Arrow indicates the extent of closure in this embryo. (h) Heart. (Images courtesy of Saadi Ghatan.)

(8.5–9.5 dpc) at the forebrain/midbrain boundary (closure 2) and at the rostral extremity of the neural tube (closure 3) (Figure 5.17). The tube then zips up caudally from closure 3 and in both directions from closures 1 and 2, eventually reaching the posterior extremity of the neural tube at the posterior neuropore. Failure or disruption of neural tube closure can have many causes, both intrinsic and extrinsic to the neural tube. The process is also a delicate one and the closure points can reopen if the embryo is handled roughly or sits in the dissection dish for too long, so be aware of these artifacts.

Following this detailed gross morphological examination, the embryos can then be used for other purposes such as looking at marker gene expression using whole-mount RNA in situ hybridization or immunohistochemistry, provided that they have been fixed and treated properly (see Section 5.D.2.b). But before getting on to that, a higher-resolution morphological assessment using histological sections and H&E staining is in order. If the time of death is between 4.5 and, say, 8.5 dpc, it is easiest to section the embryos while they are still in the uterus (see Box 5.8). Furthermore, keeping the embryos and membranes intact is especially important if you suspect or know of a problem with the fetal membranes or placental development. At later stages, the embryos can be fixed, embedded, and sectioned after they are dissected out of the uterus (see Box 5.12), but extraembryonic structures should also be processed so as not to lose information about the membranes and placenta.

Fixing and embedding later-stage embryos within the intact uterus is also an option, even though assigning a genotype will be more difficult (see below). The importance of collecting

Box 5.12 Orientation of Embryos for Histological Sectioning: Dissected Embryos

For the interpretation of histological sections, it is critical to know the orientation of the embryo and the plane of section, and nowhere is this more important than when comparing mutant and wild-type embryos. Standard orientation planes for sectioning are sagittal, transverse, and frontal (Figure 5.18A), which all depend on the position of the embryo in the block of paraffin during embedding (MM3). The following guide provides one method for achieving these orientations in sections of embryos that have been dissected out of the uterus and away from the placenta and extraembryonic membranes. For embryos still within the uterus, see Box 5.8. Variations are possible as long as you are consistent. Start by labeling the left side of the embedding mold.

- *Sagittal sections.* Lay the embryo on its right side in the mold with its head to the left, toward the label. Sections will start on the right side, because the deepest aspect of the mold will be the front of the paraffin block (Figure 5.18B).
- *Transverse sections.* Position the embryo head down with its back to the label on the left side of the mold. Sections will start at the top of the head. On the slides, the right side of the embryo will be on the left (Figure 5.18C).
- *Frontal orientation.* Place the embryo face down with the head toward the left of the mold. The front of the embryo will be sectioned first. On the slides, the right side of the embryo will be on the left (Figure 5.18D).
- When sectioning the embryos, mount the block on the microtome with the label on the left. Trim the block, placing a notch on the upper right-hand corner to maintain the orientation of the sections. Mount on labeled slides as shown (Figure 5.18E), keeping the shiny side of the ribbon down.

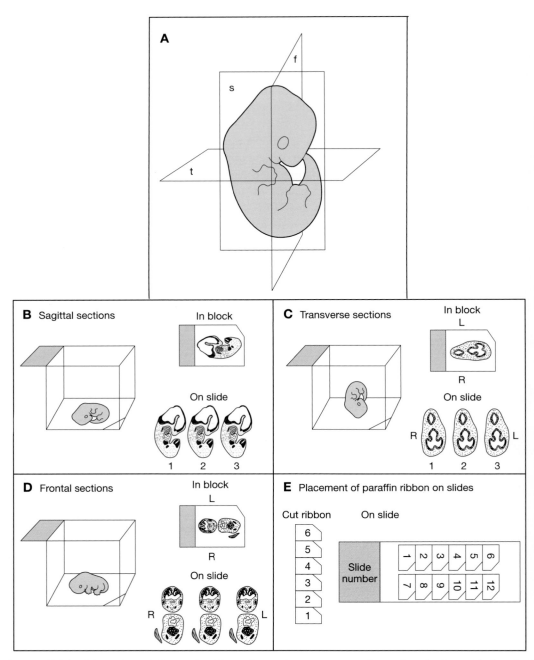

Figure 5.18. Orientation of dissected midgestation to late-gestation embryos for histology. (*A*) Standard sagittal (s), transverse (t), and frontal (f) planes of sectioning. After fixation and clearing the embryo, label an embedding mold on the left-hand side and position the embryos as shown for sagittal sections (*B*), transverse sections (*C*), or frontal sections (*D*). For sectioning, mount the paraffin block on the microtome with the label on the left and trim the block, putting a notch in the top right-hand corner for orientation. Mounting the ribbons on labeled slides as shown (*E*) will result in the embryo orientations indicated in *B–D*. Several rows of ribbon can be mounted on each slide if they are narrow enough. The numbers indicate the order of serial embryo sections. (R) Right side of the embryo; (L) left side of embryo; (1–12) the order of sections.

serial sections with a known orientation cannot be overemphasized. In our experience, it is rare to find a histological service facility that can adequately prepare histological sections from embryos, so it is probably best to do it yourself, particularly if you are concerned about the plane of sectioning. The main source for comparison and analysis of histological morphology is the excellent book, *The Atlas of Mouse Development* by Matt Kaufmann (see Appendix 1); refer to that book to see what you are aiming for.

A good morphological assessment of a serially sectioned embryo will provide an immense amount of information about the mutant phenotype. Presence and normality of tissue types, organ developmental stages, cell death, and proliferation can all be determined. As a supplement to H&E staining, specialized histochemical stains can be used to highlight specific cell types. If indicated by the mutant phenotype, scanning or transmission electron microscopy could be used to provide a more detailed morphological assessment. But the embryo is complicated, and interpretation of early embryonic stages is a skill that must be learned. Consultation with an expert or at least a few good books may be a good idea at this time (Appendix 1).

When abnormalities are detected, care must be taken in attributing them to the mutant genotype, because abnormalities are frequently observed in nonmutant embryos. If you collected the yolk sac or a limb for genotyping, you will be able to make a positive identification of the mutants. But if you sectioned embryos in utero, you will not know the genotypes. One option is to examine enough embryos to trust that a Mendelian ratio of phenotypes reflects the Mendelian ratio of genotypes, but you will always have to describe the abnormal embryos as "putative mutants" and accept that the occasional abnormal but nonmutant embryo may confound your determination of the phenotype. (If the phenotype of the homozygous mutants was well characterized on the basis of gross morphology of genotyped embryos, it may not matter too much if you do not genotype embryos from histological sections.) Alternatively, you can genotype the embryos from sections after you have finished the morphological analysis. Basically, this involves removing the coverslips, or using a slide that has not been stained and coverslipped, and collecting a sample of the embryo section for PCR analysis, carefully avoiding contamination with maternal tissue. It is possible to scrape away the maternal tissue with a razor blade or scalpel under a dissecting microscope and extract DNA for PCR analysis from the isolated embryo section using cloning rings placed directly on the slide over the section. A more sophisticated, and probably more reliable, method is laser-assisted microdissection, or laser capture, in which a defined area of the section is laser dissected and "captured" in a tube for PCR. The only problem might be the availability of laser microdissection equipment, although the technology is increasingly used in cancer research and may be available in pathology departments.

5.D.2.b Molecular characterization of mutant phenotypes

Analysis of the morphology of an embryo, placenta, or fetal membranes will provide a good indication of the affected tissues. When this information is combined with information on the normal gene expression pattern of the mutated gene during the period at which the abnormality first appears, it may suggest a hypothesis regarding the developmental role of the gene. One of the most popular and effective means of testing such hypotheses is the molecular characterization of the mutant embryos to determine patterns of gene and protein expression. For this analysis, a variety of molecular markers of differentiation is used to determine precisely which

> **Box 5.13** Analysis of Cell-type-specific Gene Expression
>
> There are many options for detecting gene expression in specific cell types or tissues at the level of mRNA, protein, or protein activity. Whole embryos or partially dissected fetal tissue (i.e., fetal organs) can be processed intact for nonradioactive mRNA in situ hybridization, immunohistochemistry, or histochemistry (e.g., β-galactosidase or alkaline phosphatase activity) to see the overall pattern of expression. These whole mounts can be subsequently sectioned and examined with or without counterstaining for a more detailed histological picture. DIC microscopy can provide tissue structure information, obviating the need for a counterstain. If you do use a counterstain, choose one that does not obscure the gene expression signal.
>
> Alternatively, embryos or tissues can be fixed and sectioned first and then processed for radioactive or nonradioactive mRNA in situ hybridization, chromogenic or fluorescent immunohistochemistry (standard or confocal microscopy), or histochemistry. Sections are especially advantageous when embryos or tissues are too large for reagents to penetrate efficiently, causing high background or trapping of reagents in cavities of whole mounts. Sections of varying thicknesses can be cut for different purposes from tissues that have been embedded in different media: Plastic media are used for very thin sections (0.2–2 µm), paraffin wax is used for sections typically between 4 and 8 µm, OCT compound (Sakura Finetechnical Co., Japan) is used for cryosectioning at 6–20 µm, and agarose can be used to embed organs such as the brain to cut 10- to 200-µm-thick sections on a vibratome. Also, metal or acrylic blocks called brain matrices are specifically designed to hold a mouse brain, allowing you to make 1-mm sagittal or coronal brain slices using razor blades. These matrices could be used with other tissues as well.
>
> You are not limited to detecting a single mRNA or protein per sample. mRNA and protein can be co-localized by sequential colorimetric in situ hybridization and immunohistochemistry (or histochemistry in the case of β-galactosidase) with different substrates to produce different color precipitates. Similarly, two different mRNAs can be detected using probes that are differentially labeled.
>
> The following are a few other helpful hints:
>
> - When sectioning, mount a fairly small number of sections on each slide so that you have plenty of slides for controls.
> - Use alternate slides for controls to have comparable sections.
> - For mRNA detection, be sure to work under RNase-free conditions at all times, even during initial dissection.
> - Nonradioactive mRNA in situ hybridization is particularly convenient and equally sensitive as radioactive in situ hybridization in most cases.
> - In whole-mount embryos, open large cavities such as the neural tube and peritoneum to prevent trapping and assist penetration of reagents.
>
> For standard methods for mRNA in situ hybridization or immunohistochemistry, see MM3 and references in Appendix 1.

tissues have differentiated molecularly, as well as the nature of any abnormal structures. Many gene products provide markers of specific tissues, particular stages of development, or particular cell states at different times in development. Judicious selection of markers appropriate to the abnormality observed will give your gross morphological and histological assessment the added strength of molecular backup. Two markers per affected tissue are usually adequate, provided that you have chosen them well to address the specific abnormality. Be aware that the absence of a marker could be the result of a lack differentiation within the tissue or a lack of the tissue itself. In addition, you may be able to pinpoint specific signaling pathways that are disrupted in the mutant embryo by the pattern of genes expressed.

Most commonly, probes for mRNA are used to detect molecular markers, although antibodies are also very useful and provide slightly different information (i.e., the location of the protein, rather than the message). Both can be used on histological and cryostat sections, or on whole, dissected embryos (referred to as whole mounts), which can be sectioned on a cryostat or vibratome after initial observation, to provide temporal and spatial gene expression pattern data that will be informative when mutants are compared with controls (Box 5.13).

5.D.2.c Cell proliferation and cell death

Embryonic growth and development are not only the result of rapid cell division but also cell migration, regional differences in mitotic rate, as well as specific sites of programmed cell death (apoptosis). Mutant phenotypes frequently include alterations in proliferation or apoptosis, either as a primary result of the mutation or as the result of secondary effects on embryo patterning or viability. Provided that you have stage- and age-matched normal controls for comparison, the assessment of proliferation and cell death in mutant embryos can contribute to an explanation of the phenotype. You may wish to analyze the entire embryo or concentrate on a specific area of expression of your gene or a structure that shows a phenotype. These two measures are generally complementary in the analysis of a mutant phenotype. For example, a mutant phenotype characterized by the underdevelopment of some structure could result from hypoproliferation, failure of cell migration, and/or an increase in cell death. Likewise, an apparent overgrowth of a structure might be caused by hyperproliferation, or more rarely, hypertrophy, or by a lack of apoptosis (assuming that apoptosis is normally part of the development of the structure). Because apoptosis is usually restricted to very specific areas in the normal embryo, elevated levels can indicate a generalized detrimental effect (e.g., vascular insufficiency leading to cell death) or specific effects (e.g., aberrant signaling triggering the apoptotic response in a specific tissue). Many methods can detect proliferation or cell death, and they can be applied to whole embryos or sections, depending on embryonic age and level of detail desired.

Measuring cell proliferation. Because embryonic cells divide rapidly, the problems associated with obtaining measurable mitotic indices from a slow-growing tissue do not apply. Thus, although it is possible to administer an extrinsic mitotic label such as bromodeoxyuridine (BrdU) to an embryo over a period of time by administering it to the pregnant mother, or to inject the mother with colcemid to arrest and accumulate mitoses in the embryo, this is usually not necessary. Intrinsic markers of mitosis are adequate, and it is nice to be able to avoid problems associated with hazardous chemicals. The simplest method to quantitate proliferation is to assess cell division in H&E-stained histological sections. Different tissue types can be easily distinguished and mitoses are clearly evident by the densely stained chromosomes in mitotic figures. The high mitotic index of the ependymal cells lining the lumen of the brain and spinal cord provides a clear example (Figure 5.19A). Because examination of histological sections is frequently part of the analysis of a mutant phenotype anyway, this is often the easiest and most convenient way to assess whether additional analysis is necessary. (Incidentally, cell death can also be recognized by nuclear morphology in histological sections by the presence of darkly staining pycnotic nuclei, so this also provides an assessment of possible alterations in apoptosis; see the next section.)

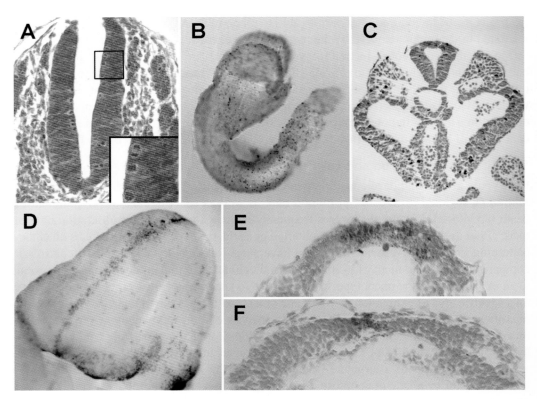

Figure 5.19. Detection of cell proliferation and cell death. (*A*) H&E staining of a section of the neural tube of a 10.5-dpc embryo. Numerous, darkly staining mitotic nuclei are visible at the luminal surface of the neuroepithelium. The boxed area shown at higher magnification in the inset shows a nucleus in anaphase and two nuclei in metaphase. (*B*) Whole mount of an 8.5-dpc embryo with mitotic cells shown by immunolocalization of phosphohistone H3. (*C*) Transverse section through the midgut region of a 9.5-dpc embryo with mitotic cells visualized by immunolocalization of phosphohistone H3. Immunolocalization was done on the whole embryo, which was then embedded, sectioned, and counterstained with Nuclear Fast Red. (*D*) Detection of cell death in the head region of a 9.5-dpc embryo using the terminal dUTP nick end labeling (TUNEL) assay on a whole mount. The midline streak of apoptotic cells marks the point of closure of the neural tube; cell death is a normal part of this process. (*E*,*F*) Detection of apoptotic cells at the site of neural tube closure by immunolocalization of activated caspase in sections of the rhombencephalon and prosencephalon, respectively, counterstained with Methyl Green. (*B*,*C* courtesy of Amalene Cooper-Morgan and Virginia Papaioannou; *D*–*F* courtesy of Saadi Ghatan.)

Antibodies to proteins involved in the mitotic cycle offer another convenient means of visualizing and quantitating cell proliferation, both in sections and in whole embryos. When you are first working out the protocols, sections of adult intestinal epithelium provide a convenient, rapidly dividing control tissue. For the analysis, always compare homozygous mutants with wild-type embryos of the same developmental stage. Antibodies to proliferating nuclear cell antigen (PCNA) are commercially available and often used in histochemical staining procedures on sectioned tissue to classify cells into phases of the mitotic cycle. PCNA is present in all mitotic stages, but is differentially expressed during different phases of the cell cycle such that G_1, S, G_2, and M can be recognized. However, this requires interpretation of staining density and may be a bit more elaborate than necessary for an initial survey of cell proliferation in mutant embryos, unless, of course, you suspect that an effect on cell cycle progression could be part of the phenotype.

Immunohistochemical staining with antiphosphohistone H3 antibody, which characterizes mitosis (and meiosis) and is also commercially available, offers a suitable marker for embryonic cell proliferation and can be applied to early whole embryos or to sections (Figure 5.19B,C). The advantage of doing whole mounts, of course, is that overall patterns can be seen in the intact embryo. However, the limitation is the depth of penetration of the reagents, so this procedure works best on young embryos (≤10.5 dpc) or on dissected tissues from older embryos. Whole embryos can later be embedded and sectioned for detailed cell counts and determination of the mitotic index (MI) once the overall patterns have been assessed. A counterstain such as Nuclear Fast Red will be necessary so that the MI can be calculated and compared (MI = number of mitotic cells/total number of cells examined). The MI can be calculated for specific tissues or for the embryo as a whole, provided that the total number of cells examined is reasonably large (several thousand cells).

Measuring cell death. Alterations in tissue dynamics may involve changes in both proliferation and cell death and so assays for both are often considered together. Apoptosis is a normal, reproducible part of embryogenesis and may be disturbed in a mutant phenotype. Necrotic cell death may also be part of a mutant phenotype if placental or vascular insufficiency leads to tissue necrosis. Not all assays of cell death allow you to distinguish between apoptosis and necrosis. Apoptosis can be recognized morphologically by characteristic features in H&E sections that include condensation of chromatin (sometimes in association with the nuclear membrane), nuclear fragmentation, blebbing, and fragmentation of cytoplasm into apoptotic bodies, often surrounded by a clear halo. Apoptotic indices can be calculated in the same way as mitotic indices, and histological sections will sometimes provide adequate information to rule out (or rule in) apoptosis as a factor in a mutant phenotype.

Alternatively, specific markers of apoptotic or dying cells can be used for quantitation. The commonly used terminal dUTP nick end labeling (TUNEL) assay is commercially available as a kit, but the method has limitations involving tissue penetration in embryo whole mounts and does not distinguish between apoptotic and necrotic cell death. Several alternative apoptotic markers, such as activated caspase, can be detected immunohistochemically with good reproducibility. A convenient test tissue for working out the method and to use as a control is a 12.5-dpc limb bud (either whole mount or sections) because apoptosis takes place in the interdigital web tissue. In addition, several places in the embryo in which apoptosis occurs normally (e.g., along the ridges of the closing neural tubes at 9.5 dpc [Figure 5.19D–F], in the dorsal root ganglia at 12.5 dpc, and in the central region of the presumptive retina at 11.5 dpc) can be used as internal controls. As with measurements of cell proliferation, both the pattern of apoptosis and the frequency as measured by the apoptotic index can provide information on mutant phenotypes.

The histological features of necrotic cell death closely resemble apoptosis. However, necrosis generally begins in the embryo in foci that expand to encompass large areas. If this pattern of cell death occurs in the homozygous mutant embryos, you might have recovered mushy, degenerating remains, which could not be processed for histology. A necrotic embryo would immediately suggest placental or circulatory insufficiency, and thus, generalized embryonic cell death may be secondary to a developmental or functional problem with the extraembryonic structures or the circulatory system. In this case, the development of the placenta and all of the extraembryonic structures should be examined.

5.D.2.d Embryo, tissue, and cell developmental potential

In a developing organism, the normality of the whole depends on the correct functioning of all of the parts. A defect in one cell type or tissue of an embryo will quickly compromise neighboring or dependent cells, setting up a chain reaction of problems that can encompass the entire organism. This effect is particularly marked in defects that have their onset during early development. Our goal in pinning down the time of death and then trying to determine the earliest onset of the mutant phenotype is to identify the earliest role of the gene that starts the catastrophic cascade. However, the effects of an early phenotype may be just the tip of the iceberg in terms of gene function. You may know that your favorite gene is expressed more widely in time or space than the earliest-identified defective tissue. Or you may not be able to distinguish a primary defect due to the mutant gene from a secondary defect due to the accumulation of problems in a developmental cascade. If your primary goal is to study the role of your gene in later embryonic stages or even adult tissue, you may be able to entirely circumvent early embryonic lethality to get at these stages. To give it a try, go to Chapter 8.

On the other hand, much information can be gained by short-term embryo culture, allowing you to observe development directly. Alternatively, removing specific tissues from the detrimental effects of a dying embryo and testing developmental potential either in isolation or recombined with normal tissue are useful adjuncts to studying in utero development. These approaches are useful for sorting out primary and secondary effects of the mutation, and you may also be able to get at some of the later effects of the gene in cases for which a placental insufficiency is limiting. The potential of mutant tissue or organs to survive and differentiate when removed from developmental constraints, and/or from detrimental effects due to failure of other essential organs, may provide valuable clues to later gene function that would otherwise be obscured by early lethality. What follows are some of the variety of ways to test the developmental potential of embryonic tissue. These should be selectively applied to your mutant phenotype depending on how far it develops before the lethal effect appears and on what you have already learned about the lethal effect. As always, the experiments should be guided by the expression pattern of the gene. However, even genes with ubiquitous expression can have tissue-specific defects.

In vitro culture of postimplantation embryos. Culture of preimplantation and periimplantation embryos was discussed in Section 5.C. For later-stage embryos, whole-embryo culture is feasible for periods of 24–72 hours between 5.5 and 10.5 dpc, and gastrulation and early organogenesis can be directly observed and even manipulated. A variety of techniques for static or rotating culture has been developed, with conditions dependent on the age of the embryos at the time of explantation (MM3). Although the outermost layers of the fetal membranes (giant trophoblast cells and parietal endoderm) are removed in these protocols, the embryo is cultured with the placenta and yolk sac either intact or partially dissected so that the integrity of the circulation is retained, allowing observation of the development of these extraembryonic tissues as well.

In the context of mutation analysis, whole-embryo culture allows the direct comparison of development of homozygous mutants with wild-type embryos. The protocols are quite exacting and require a considerable investment of time and energy, but can be very informative for certain types of mutations, such as those affecting morphogenetic movements. The advent and increasing availability of transgenic animals with fluorescent proteins localized to different subcellular organelles can be combined very usefully with whole-embryo culture and regular fluorescent or confocal microscopy so that the developmental event can be observed in real time in live embryos. You may wish to cross your mutant with mice carrying such fluorescent markers to label all cells, or, if you put such a fluorescent marker into your targeting construct as an expression reporter (Section 2.B.2) you can use whole-embryo culture to track the mutant (and wild-type) cells that express your gene.

In vitro culture of cells, tissues, and organs. The long history of experimental embryology has provided us with techniques for isolating almost any embryonic organ, germ layer, or cell type, and a myriad of cell and tissue culture methods have been developed for specific tissues (MM3). The isolated tissues can be grown in isolation or recombined with other, interacting tissues. For example, the early germ layers of the embryo can be separated by a combination of mechanical and enzymatic dissection for culture in isolation; the allantoic bud can be grown in organ culture or cells can be dispersed and grown as a monolayer; limb buds, lung buds, or mammary buds can be isolated and undergo development in vitro; the layers of the yolk sac can be separated and cell lines derived from them; ureteric buds can be isolated and recombined with mesenchyme and will undergo branching morphogenesis.

In the context of mutant analysis, these techniques can be used to study an affected organ or tissue for several days beyond the lethal time point of the whole embryo. Because of the inherent variability associated with tissue and organ culture, the key is to explant the tissue, after carefully stage-matching the embryos, before the onset of the mutant effect. Then, compare the behavior of mutant and wild-type tissue. For the most part, only short-term cultures will be informative because the techniques do not support normal development in the long term. You might be able to gain some time by rescuing the mutant tissue from a placental (or cardiovascular) defect and thus watch the unfolding of mutant effects in embryonic tissues in vitro. Assays of cell proliferation, death, and migration, as well as the use of molecular markers, may all be useful in combination with in vitro culture. Similar to whole-embryo culture, the use of fluorescent protein markers allows you to study dynamic cell behavior of all mutant cells or of expressing mutant cells, if you have a reporter knock in. In addition, these in vitro culture methods allow the recombination of tissues, and, depending on the type of phenotypic effect observed, it may be useful to combine mutant and wild-type tissue in organ culture to test for the integrity of inductive interactions in the mutant tissue.

Differentiation of early embryos and tissue fragments into teratomas. An alternative means of determining the differentiation and long-term histogenic potential of isolated embryonic tissues is to culture them in vivo in ectopic sites in a histocompatible host. The host organism, usually an adult mouse, serves as a culture chamber, supplying nutrients and trophic factors to the transplanted tissue. Tumors called teratomas or teratocarcinomas develop from these transplanted embryonic fragments, often containing a variety of differentiated cell and tissue types in chaotic organization. Teratocarcinomas are distinguished by the presence of undifferentiated embryonal carcinoma cells, which are transplantable stem cells. The composition of the tumors indicates the potential for histogenesis of the transplanted embryonic cells. When comparing mutant and wild-type tumors, you may, for example, discover a specific cell type missing from the mutant grafts, thus indicating a role for the gene in the differentiation of that cell

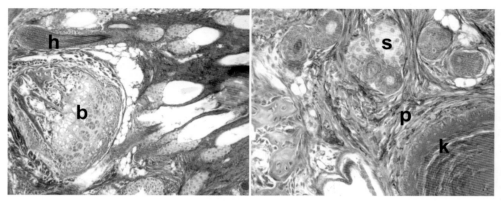

Figure 5.20. Example of a teratoma. Two sections through a teratoma show a multitude of cell types using Masson's trichrome stain. (b) Bone; (h) hair follicle; (k) keratinized epithelium; (p) pigment cells; (s) sebaceous gland.

type. One caveat in this type of analysis, however, is that the host organism could be potentially supplying a soluble factor missing from the mutant tissue as a result of the mutation. This might result in rescue of the differentiation capacity of mutant tissue but is not very informative with regard to the missing factor.

Teratomas develop from transplanted blastocysts, the embryonic portion of pregastrula- or gastrula-stage embryos, and from the isolated tail bud or genital ridges of midgestation embryos. Histocompatible hosts are essential to avoid an immune reaction. The tissue is transplanted under either the testes or kidney capsule (MM3) and left for several weeks. Analysis is usually done histologically using Masson's trichrome stain, or any number of other special histochemical stains, which more clearly delineates different tissue types than H&E staining (Figure 5.20). The tumors can also be analyzed for neoplastic potential, i.e., transplantability or the ability to grow as stem cell lines in vitro, for specific gene expression, and for the expression of specific genes. Finally, teratomas can be used as the source for making embryonic fibroblast cell lines (see next section).

Making embryonic stem, trophoblast, or embryonic fibroblast cell lines. Another means of revealing the potential of embryonic cells outside the context of the complete organism is to derive stem cells directly from the embryo (MM3). You will already be familiar with the capabilities of these cells if your mutant mice were made using embryonic stem (ES) cell technology. Now, in the analysis of mutant mice derived from that technology, you can once again make use of the amazing properties of ES cells, but this time to inform on the mutant phenotype. If, for example, your homozygous mutant dies at 6.5 dpc from a placental defect, but you know that the gene is expressed later in development in perhaps yolk sac mesoderm or nerve cells, you may wish to try deriving ES cells from homozygous mutant embryos.

If this works, you will have a source of homozygous mutant cells that can be subjected to culture conditions that promote mesoderm and/or neuronal differentiation, allowing you to study these processes in vitro. In the course of making ES cell lines from homozygous mutant embryos, wild-type and heterozygous cell lines will also be generated that will serve as controls and, possibly, material for uncovering dosage effects.

> **HELPFUL HINT**
>
> *ES cells are most easily derived from 129 strain embryos or from embryos with a large component of 129 such as B6129F1. Make every effort to work with this strain for deriving ES cells; you will not regret it. An alternative to making new ES cell lines from homozygous mutant embryos is to use the original targeted ES cells, which are heterozygous, and make them homozygous either by retargeting the wild-type allele or selecting highly resistant cells with a high drug dose (MM3).*

Although less common than ES cells, stem cells known as trophoblast stem (TS) cells can be derived from the precursors of trophoblast lineages (MM3). If your mutation affects the placenta or trophoblast, it may be worth deriving TS cells for in vitro study.

If making ES or TS cell lines from your homozygous mutant embryo seems like a good idea, stop for a moment and anticipate another potential use for these cells: making chimeras, as described in the next section and in Chapter 8. Before deriving ES or TS cells from the mutant embryos, consider breeding in a transgenic cell marker by breeding heterozygotes with a marker-bearing strain to make the best use of the cells in chimeras (see Box 8.1, p. 180).

Finally, primary embryonic fibroblast cell lines, called mouse embryonic fibroblasts (MEFs), can be made from embryos (MM3) or even teratomas. These may be useful for comparing proliferation rates or ease of immortalization between homozygous mutant and wild-type fibroblast cells.

Testing developmental potential in chimeras. Even if your homozygous mutant embryos die by 9.5 dpc, it may still be possible to examine mutant cells in the context of a developing embryo at later stages using combinations of wild-type and mutant tissue. The idea is to recover some mutant cells before the time of homozygous mutant embryo lethality and combine them with wild-type cells in a chimera. Provided that you have a cell-specific marker to distinguish the mutant and wild-type cells (Box 8.1, p. 180), you can follow the cells during development to determine whether the mutant cells behave normally or aberrantly. This procedure is sometimes referred to as chimeric rescue, although this is a bit of a misnomer. Chimeras might successfully circumvent lethality, leading to rescue of the mutant cells or even the embryo, but just as likely, the mutant cells may still suffer cell-autonomous detrimental mutant effects. Nonetheless, much information can be obtained by examining the mutant cells' behavior in combination with wild-type cells. Because making chimeras can entail a lot of work, think carefully about the questions that can be answered by this approach. You will only want to consider this approach once you have a very clear idea of the cause of lethality. Then, if the homozygous mutants die early and you know that the gene is expressed later in tissues that you wish to study, or if you want to know whether a specific mutant phenotype is cell autonomous or not, go to Chapter 8.

5.E Analyzing Midgestation to Late-gestation Phenotypes

Midgestation to late gestation in the mouse is characterized by a period of tissue differentiation and maturation, organogenesis, and fetal growth. Thus, many of the mutant phenotypes that are first found during this period of development include those that affect embryo viability and organogenesis. Not all phenotypes that have their origin during late gestation will be lethal and many will be evident as birth defects in viable animals. The techniques for detection and analysis are similar and you may have been sent to this section from Chapter 6 if you have viable mutants with postnatal defects.

Mutant embryos die after 9.5 dpc	Go to Section 5.E.1 ▼
Postnatal phenotype indicates origin of an abnormality during late gestation	Go to Section 5.E.2 ▶

5.E.1   Analyzing phenotypes lethal after 9.5 dpc

If an embryo is present at the 12.5-dpc dissection point, but is dead and possibly degenerating, death sometime after the establishment of the chorioallantoic placenta is indicated. With the fusion of the allantois to the chorion and the formation of umbilical vessels within the allantois, the chorioallantoic placenta takes over the jobs of supplying the embryo with nutrients and oxygen, as well as waste removal, functions that up to this point were satisfied by diffusion across the fetal membranes. At 12.5 dpc, the liver takes over from the yolk sac blood islands in producing blood, and the heart pumps blood throughout the rapidly growing embryo and placenta. Consequently, common causes of embryonic death during this period are cardiovascular, hematopoietic, or placental insufficiency.

Any fetuses that die this late will not be completely resorbed and will be born with the rest of the litter. However, they will likely be eaten along with the placentas before you have a chance to see them, assuming that you work regular hours. If you happen to find dead, developmentally retarded pups in the cage soon after parturition, it could indicate death before birth, as opposed to perinatal death. Even if the pups died perinatally, you may find yourself back in this section to trace developmental defects that had their origin earlier in development.

If fetuses survive to late fetal stages before succumbing, most of their development is available for study. However, there is also more time for early, nonlethal mutant phenotypes to become compounded through a developmental cascade of secondary and tertiary effects. Just because an embryo dies late does not mean that the mutant effects occur late, so determining the time of death is only a first step. In fact, the immediate cause of death may be accompanied by many developmental abnormalities that had their origin much earlier but were not life threatening. During these late fetal stages, the essential organ systems are the heart, circulation, blood-forming tissues, and placenta. Abnormalities in virtually all other organ systems can be tolerated at least until birth.

5.E.1.a Hematopoietic defects

The circulation of blood cells, specifically, the oxygen-carrying red blood cells, is essential for embryonic growth and development. Thus, hematopoietic defects that cause embryonic lethality after 9.5 dpc are most likely those that compromise red blood cell formation or integrity. A deficiency of functional red blood cells will create an oxygen deficit, leading to tissue growth retardation, developmental delay, and, if severe, death of the embryo. Embryos with hematopoietic defects will usually appear pale because of anemia and perhaps growth retarded in comparison to controls.

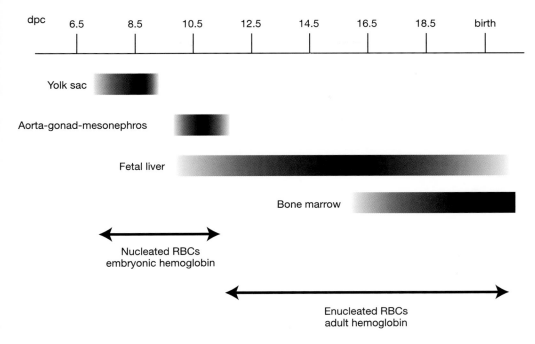

Figure 5.21. Developmental timing, tissue distribution, and erythrocyte characteristics during blood cell formation in the mouse embryo. Shaded bars indicate the time period at which hematopoiesis occurs in the designated tissue. The intensity of shading indicates the relative amounts of hematopoiesis in the tissue. Arrows indicate the timing of formation of nucleated erythrocytes (red blood cells, RBCs) expressing embryonic hemoglobins, or enucleated erythrocytes expressing adult-type hemoglobins. (Modified, with permission, from Baron, *Exp. Hematol. 31:* 1160–1169 [©2003 International Society for Experimental Hematology].)

Hematopoiesis occurs at different locations within the mouse embryo at different stages of development (Figure 5.21). Blood cell formation initially takes place in the blood islands of the yolk sac starting at 7.5 dpc and continues until 9.0 dpc. In contrast to the enucleated red blood cells formed at other sites, these yolk-sac-derived erythrocytes retain their nuclei. At 9.5 dpc, sites of hematopoiesis are found in the embryo, namely, in the aorta-gonad-mesonephros (AGM) region and fetal liver, which remain prominent sites of fetal hematopoiesis until just after birth. By 15.5 dpc, hematopoiesis initiates in the bone marrow and spleen, which remain sites of hematopoiesis throughout adult life. The erythrocytes generated by the fetal liver, bone marrow, and spleen are enucleated. The type of hemoglobin expressed by the red blood cells also changes during development (Figure 5.21), each with its own oxygen-carrying characteristics. Blood cell formation is a collaboration between the blood-forming progenitor cells (Figure 5.22), the stromal microenvironment, or growth factors produced in other tissues (e.g., erythropoietin production from the kidneys). Thus, defects in any of these tissues can result in red blood cell mutant phenotypes.

You may have been expecting a hematopoietic defect based on your knowledge of the protein encoded by your gene of interest and/or its expression pattern. However, if the protein is novel or the expression pattern uninformative (e.g., ubiquitous expression), then you will likely be considering a hematopoietic defect if you noticed that your homozygous mutants, includ-

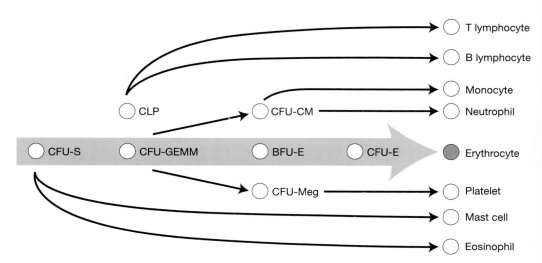

Figure 5.22. Red blood cell development. The pluripotent hematopoietic stem cells differentiate into multipotent and unipotent progenitors with distinct differentiation potentials. The erythrocytes are derived from the multipotent CFU-GEMM (colony-forming unit–granulocyte, erythroid, macrophage, megakaryocyte) that gives rise to unipotent erythroid progenitors, burst-forming unit–erythroid (BFU-E), and colony-forming unit–erythroid (CFU-E). (CFU-S) Colony-forming unit–spleen; (CLP) common lymphoid progenitor; (CFU-GM) colony-forming unit–granulocyte, macrophage; (CFU-Meg) colony-forming unit–megakaryocyte. (Modified from Speck et al., *Mouse development: Patterning, morphogenesis, and organogenesis* [ed. J. Rossant and P.P.L. Tam], pp. 191–210, Academic Press, San Diego, California [©2002 Elsevier].)

ing their livers, appear pale and growth retarded. A very simple first step in the analysis of a suspected hematopoietic defect is to generate a blood smear or cytospin preparation of the circulating blood to examine the morphology and staining characteristics of the blood cells and whether they are nucleated. Abnormal morphology may indicate a defect in formation that could cause the mutant red blood cells to be more fragile or less hemodynamic for transit through the vasculature. With sufficient numbers of blood cells, you may also be able to determine the hemoglobin concentration per cell. Reduction in the hemoglobin concentration in the red blood cells will result in a lowered ability to carry oxygen to the tissues. You can also examine the type of hemoglobin (embryonic or adult) contained within the homozygous mutant red blood cells by protein or mRNA expression analysis.

A blood smear will provide an initial piece of evidence to indicate a hematopoietic defect. Next, determine when the homozygous mutant phenotype first appears during development. Analysis of the gross morphology and histology of the homozygous mutant embryos in comparison to controls is the subsequent step in the analysis. Pay particular attention to the red- blood-cell-forming tissues, noting a presence or deficiency of blood cells. If there appears to be a reduction in the blood-cell-forming tissues, then the red blood cell progenitors can be examined. Some of these progenitors can be grown in vitro in semisolid media where they will give rise to colonies of erythrocytes (i.e., a clonogenic assay) and thus can be quantified (e.g., number of colony-forming units–erythroid or CFU-E, number of burst-forming units–erythroid or BFU-E) (Figure 5.22). It is also possible that even more primitive multipotent hematopoietic stem cells may be defective, and these can also be examined in vitro using clonogenic assays. Subsequent molecular marker, cell proliferation, and cell death analyses (see Section 5.D.2.c) can also be used to further define the tissue and stage at which the mutation acts.

Other avenues of analysis of the mutation include a chimera analysis (Chapter 8), in vitro differentiation of homozygous mutant ES cells into blood cells (Section 5.D.2.d), and, depending on the expression pattern, conditional knockouts in specific tissues (Chapter 8) to determine which defective tissue is causing the hematopoietic abnormality.

5.E.1.b Cardiovascular problems

Cardiovascular problems are a major cause of postimplantation lethality. Failure in heart specification or early differentiation can cause lethality before 9.5 dpc. After that time, insufficient cardiac function, structural defects, or conduction system defects can result in late-gestation death. From an early stage, the cardiovascular system depends on all of the various components developing and working in synchrony. A problem in one area can rapidly lead to a cascade of

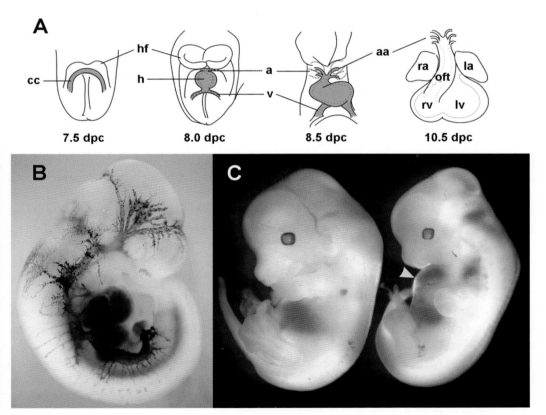

Figure 5.23. Cardiovascular development. (*A*) Ventral views of stages of heart development between 7.5 and 10.5 dpc. Cardiomyocytes first appear in the cardiac crescent (cc) at the headfold (hf) stage. The endocardial tubes fuse in the ventral midline to form the tubular heart (h) with an arterial and venous (a and v, respectively) pole. This tube loops into a C shape and then forms the four chambers of the heart: the right atrium (ra), left atrium (la), right ventricle (rv), and left ventricle (lv). At the arterial pole of the heart, the outflow tract (oft) connects with the aortic arch (aa) arteries in the pharyngeal region. (*B*) India ink injection into the heart of a 10.5-dpc embryo, viewed from the right side. The ink particles fill the heart and outline vessels throughout the body. (*C*) A wild-type 12.5-dpc embryo on the left and a *Tbx2* homozygous mutant embryo on the right. The mutant embryo displays cardiac effusion or a swollen pericardial sac (*yellow arrowhead*). (*A,B* Courtesy of Robert Kelly; *C* provided by Zach Harrelson and Virginia Papaioannou.)

secondary morphological and functional defects. Thus, when investigating cardiovascular defects, not only the heart but also the yolk sac, fetal, and umbilical vasculature, as well as the endoderm and cardiac neural crest, must all be considered. Several telltale indicators of cardiac insufficiency will be evident on dissection of embryos during the second half of gestation: growth retardation, a blood-engorged liver, tissue edema, and pericardial effusion, evidenced by a swollen, fluid-filled pericardial sac (Figure 5.23C). This pericardial edema can be the result of local problems in the heart or other defects, such as failed vascular connections, in more distal sites that lead to fluid buildup in the heart and pericardial cavity. In addition, alterations in the looping pattern of the heart (Figure 5.23A) as well as the lack or irregularity of a heartbeat can be detected at dissection. Lack of a heartbeat will quickly lead to widespread cell death in other parts of the embryo, including the vasculature. In these cases, knowing the wild-type gene expression pattern of the mutated gene will help identify the site of the primary defect.

Vasculogenesis takes place both within the extraembryonic tissues, starting as early as 7.0 dpc in the yolk sac, and a little later within the embryo, in the heart, dorsal aorta, and branchial arches. The vessels gradually coalesce into a complete circulatory system connecting the embryo with the yolk sac through the vitelline circulation and with the placenta through the umbilical circulation. Nucleated primitive red blood cells, first formed in the yolk sac blood islands, will circulate throughout. The development of vasculature depends critically on the function of the heart, not only for the delivery of oxygen to the tissues but also through hemodynamic forces that affect vascular morphogenesis. Lethality that occurs after fusion of the allantois to the chorion could result from failure of blood formation or from vascular defects, possibly, but not necessarily, associated with cardiac insufficiency. Once the time of death has been established and cardiovascular involvement is indicated, collect embryos at earlier stages and examine the gross morphology of the hearts for size, the presence of differentiated chambers, vascular connections, and laterality. Also, examine the extraembryonic membranes for the vascular pattern and presence of blood, noting any areas of bleeding or effusion of blood into tissues. To distinguish primary from secondary effects, it will be necessary to determine the earliest morphological abnormality. A simple means of visualizing the vasculature of midgestation to late-gestation embryos is to inject India ink either into the heart or a branch of the vitelline vein of the yolk sac, preferably while the heart is still beating (MM3) (Figure 5.23B; also see Figure 7.4F–H). Alternatively, the great vessels or the vasculature of the placenta or other organs can be detected by injecting with liquid plastic that solidifies to form a permanent cast of the vessels (MM3).

Observations of the external morphology of the heart and vasculature should be followed up by serial section histology, either of the intact embryo or isolated hearts. The advantage of sectioning the intact embryo, including the yolk sac, is that vascular connections and other areas of the embryo can be examined at the same time. Compare the homozygous mutant embryos with wild-type, stage-, or age-matched controls to determine if the four heart chambers are differentiating correctly and the endocardial cushions and septa are forming properly to separate the chambers. During later development, the cardiac valves that form from the endocardial cushions are necessary to ensure unidirectional blood flow and that the cardiac septa separate the right from left circulation (see Figure 6.1). The ventricles are normally completely separated by the ventricular septum by 13.5 dpc, and ventricular septal defects are a common observation in cases of cardiac insufficiency. Nucleated erythrocytes should be evident in the heart and blood vessels from about 8.5 dpc (see Section 5.E.1.a). In addition to doing routine histology, a molecular analysis should be undertaken using molecular markers, in whole mounts and sections, that are specific for different developmental stages or indicators of the differentiation of the vasculature or specific chambers of the developing heart. Alterations in cell death or cell proliferation can play an important part in cardiac abnormalities (see Section 5.D.2.c). In addition, physiological measures such as calcium flux, electrocardiograms, etc., can be taken. Sophisticated new applications of technologies such as ultrasound make it possible to visualize embryos in utero. Although not in widespread use, these technologies make it possible to analyze blood flow, heartbeat, and morphology without disturbing the embryos.

5.E.1.c Placental insufficiency

Another major cause of death after 9.5 dpc is placental insufficiency. You have reached this section because you have determined that your homozygous mutant embryos die after fusion of the allantois to the chorion. The placental abnormalities that lead to embryo lethality after this event are primarily those of the yolk sac placenta or chorioallantoic placenta, for simplicity called the yolk sac and placenta, respectively. Defects in the parietal yolk sac, which consists of trophectoderm and parietal endoderm, could also cause lethality but most likely earlier than this time point. Considering the essential nature of the yolk sac and placenta for the development of the mouse embryo to term, it is surprising how often these tissues are overlooked in expression studies. Thus, if you have identified a homozygous mutant lethal phenotype after 9.5 dpc, it is a good idea to determine if your gene of interest is expressed in the yolk sac and/or placenta, if you have not already done so. Yolk sac defects are generally easy to identify because this is one of the prominent extraembryonic membranes that you must dissect to reveal the embryo. Often, yolk sac abnormalities cause alterations in the yolk sac vasculature that are evident by simple visual inspection (e.g., a pale yolk sac). In contrast, the placenta is usually discarded without examination. Thus, at least initially, the placenta is often overlooked as a primary defect when examining an embryonic lethal phenotype.

The visceral yolk sac is a bilaminar structure that envelops the developing embryo. The two tissue layers are of distinct tissue origins: The outer endoderm layer is derived from the visceral endoderm and the inner mesoderm layer from the extraembryonic mesoderm generated by the primitive streak. Common gross mutant phenotypes associated with the yolk sac include a thin and transparent appearance or a pale appearance with an apparent absence of vasculature/blood, or an abnormal vascular pattern. The circulating blood cells in the yolk sac provide a visual marker of the yolk sac vasculature. However, if a blood flow problem causes a lack of blood cells in the yolk sac, then determining the presence of yolk sac vasculature defects requires more study. Histological analysis provides useful information with regard to the presence of the vasculature in the yolk sac. In addition, whole-mount immunostaining or in situ hybridization of the yolk sac using a vascular marker such as platelet endothelial cell adhesion molecule (PECAM) is also useful for determining the presence, extent, and pattern of yolk sac vasculature. The two layers of the yolk sac can be physically separated using enzyme treatments (MM3), providing a means to analyze each tissue in isolation. A note of caution: Yolk sac abnormalities can be intrinsic to the yolk sac tissues or secondary to cardiac defects.

ES cell-derived chimeras provide a very useful system to determine if a yolk sac defect is caused by primary abnormalities in the endoderm- or mesoderm-derived tissues. This is because one can generate polarized chimeras in which the yolk sac endoderm is either entirely mutant or entirely wild type. The yolk sac phenotypes of such chimeras can be analyzed to define which tissue is altered by your mutation (Chapter 8).

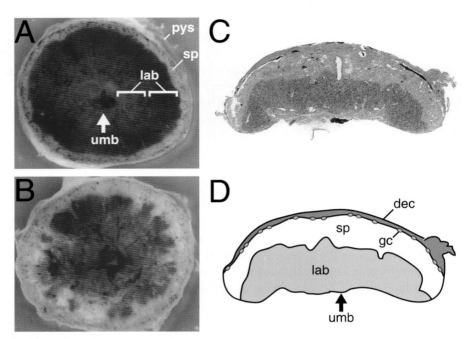

Figure 5.24. Placental morphology and histology. (*A*) An 18.5-dpc wild-type placenta viewed flat from the fetal side, showing the concentric tissue organization from the central connection of the umbilical vessels. The brackets indicate the two areas of the labyrinth inside and outside the ring of attachment of the yolk sac to the placenta. (*B*) An 18.5-dpc mutant placenta, showing disruption of tissue organization and heterogeneous morphology within the labyrinth. (*C*) Cross section of 18.5-dpc wild-type placenta, showing various tissue layers of the organ; H&E staining. (*D*) Diagram of placenta shown in *C*. (dec) Maternal decidua; (gc) giant trophoblast cells; (lab) labyrinthine layer; (pys) parietal yolk sac; (sp) spongiotrophoblast; (umb) site of attachment of the umbilicus. (*A–C* provided by Laurel Fohn and Richard Behringer.)

The placenta is composed of different tissue types of both maternal and fetal origin (Figure 5.24). From the maternal side moving to the fetal side are the maternally derived decidua, the embryo-derived trophoblast giant cells, spongiotrophoblast, labyrinth, and umbilical blood vessels. The trophoblast giant cells spongiotrophoblast is derived from the trophoblast, whereas the labyrinth is a composite tissue derived from trophoblast and extraembryonic mesoderm. The labyrinth is the site of maternal–fetal exchange. The umbilical blood vessels are derived from the allantois, an extraembryonic mesoderm derivative. Defects in any of these placental tissues can compromise embryo growth and viability.

What types of placental phenotypes are possible? If the placenta is too small, it may not be able to fully support the growth of the fetus. Interestingly, a large placenta may also be insufficient to support fetal growth effectively. Enlarged placentas can be a secondary consequence of a primary tissue defect, and enlargement may represent a physiological response of the organ to compensate for these primary defects. Thus, alterations in placental size can change over time. Placental size differences and the corresponding functional abilities of abnormal placentas lead to alterations in fetal size and viability. Accordingly, a simple analysis would be to examine the gross morphology of the placenta and determine placental and fetal weights of the homozygous mutants in comparison to controls at different stages of development. Remember to blot excess liquid from the placentas and fetuses to obtain an accurate measure of their weights. Another simple gross analysis of the placenta is to examine it flat, showing it from the fetal side (Figure 5.24A,B). A wild-type placenta will have concentric rings of healthy-appearing tissue. An abnormal placenta might show a disruption in these concentric rings of tissue, and, because the labyrinth is highly vascularized and red in color, you may find differences in color and morphology of this region.

Likely placental phenotypes include small placenta and small fetus, normal-sized placenta and small fetus, large placenta and small fetus, and large placenta and large fetus. A hypoplastic placenta indicates a defect that cannot be compensated by increased growth. A placenta of normal size with a growth-retarded fetus may indicate a defect in the placenta that does not alter its gross morphology but compromises its function. An enlarged placenta but small fetus suggests a defect that triggers compensatory growth, but also the enlarged placenta does not function sufficiently for normal fetal growth. An enlarged placenta and large fetus might suggest a general overgrowth phenotype rather than a defect limited to the placenta. Analyze timed matings at earlier and earlier stages of development to determine when you can first detect the gross morphological changes associated with the homozygous mutants.

Histological analysis of the placenta is very informative because of its stereotypic layered structure (Figure 5.24C,D). Ultrastructural analysis (transmission electron microscopy) can also be very informative because standard light microscopy of paraffin-embedded histological sections may not provide the resolution required to see the specific details of the complex structures of the placenta, notably in the multilayered labyrinth. The three-dimensional structure of the maternal or fetal vasculature of the placenta can be examined by creating a plastic cast (MM3) that provides information about the maternal–fetal interface. The trophoblast giant cells, spongiotrophoblast, and labyrinthine layers are histologically distinct, and each layer has specific molecular markers that can be used for their identification and determination of their differentiation status. Depending on the tissue defects that you uncover, you may also wish to consider examining cell proliferation and cell death (see Section 5.D.2.c).

A placental defect can lead to embryonic death, precluding your ability to study the effects of a mutation in the fetus at later stages. A useful method to bypass defects in placental structures derived from the trophoblast is to generate chimeras using wild-type tetraploid morulae or blastocysts in combination with ES cells or diploid morulae (Chapter 8). Thus, wild-type function can be provided in the placenta by the tetraploid component of the chimera, allowing the mutation to be analyzed in isolation in the developing fetus.

In the case of inbred mouse strains, the mother and fetus are histocompatible. On the other hand, even histoincompatible fetuses develop without immune rejection by the mother. Thus, there must be mechanisms to allow growth and development of fetuses in mothers that would otherwise recognize these fetuses as foreign tissues. One can imagine that if there are defects in these mechanisms, then the mother may immunologically reject the developing histoincompatible fetuses. Though very rare, this type of mutant phenotype is possible and would be indicated by immune-based phenotypes present at the maternal–fetal interface. If you suspect such a mutant phenotype, you can test this idea by determining if the mutant phenotype can be rescued if the mother is either histocompatible or genetically immune-compromised (e.g., by using the *Foxn1*, *Prkdc*, or *Rag1* mutations).

> **Box 5.14** How Many Mutants Should Be Examined?
>
> This question applies not only to gross morphological analysis but also to cell and molecular studies. You must correlate abnormal phenotypes with a mutant genotype to conclude that the mutation leads to the development of the abnormalities. At a minimum, one would want to analyze at least three mutants and three controls for each time point examined. Of course, analyzing greater numbers of mutants will make you more confident in your understanding of the mutant phenotype. If you have a mutant phenotype with variable penetrance and expressivity, you will potentially need to analyze many more mutants to correlate the phenotype with the genotype and also to determine the incidence of the phenotype among homozygous mutants.

5.E.2 Developmental defects in morphogenesis and organogenesis

There are relatively few primary causes of death after 9.5 dpc (Sections 5.E.1.a–5.E.1.c) but many causes of morphological abnormalities, which can be associated with lethal or nonlethal mutations. You may have uncovered a morphological defect during the study of your homozygous mutants either prenatally or postnatally. You are now interested in determining the molecular, cellular, and embryological mechanisms that lead to the mutant defects. Your goal will be to determine when and in which cells or tissues the primary defect initiates the mutant phenotype that you have identified. Thus, it is essential to become well familiarized with what is known about the development of the tissue or organ in question. Accordingly, as you begin your detailed analysis, it will be useful to answer some fundamental questions to provide a framework for a detailed analysis of the mutant phenotype to identify the primary defect:

- Which tissue(s) expresses the gene product in question?
- When does the tissue or organ form during embryogenesis?
- At what stage does expression of the gene in question initiate in the tissue?
- When does the mutant phenotype first become apparent in the forming organ?
- Is the mutant phenotype completely penetrant and fully expressed or is there variability?

5.E.2.a Morphology: Gross and histological

The initial step in the analysis is to document the development of the gross morphological defect in the homozygous mutants. Analyze sufficient numbers of mutant and control embryos at different embryonic stages to determine when the defect is first apparent (Box 5.14 and see Figure 5.11). This will provide a starting point for the analysis because it indicates when the outcome of the mutation first has a detectable impact on normal development.

Examination of fetuses should include both external and internal gross morphology. Using the expression pattern of the gene as a guide, dissection of different organ systems can be very informative. Do not be afraid to disassemble the embryos to isolate relevant internal organs. The skeleton can be readily visualized intact at any stage of gestation (Box 5.15).

Histological analysis provides essential information on the microanatomical defects of the tissue or organ of interest. Depending on the situation, you may want to fix and section the embryos or fetuses within the uterus (see Box 5.8). However, with late-gestation fetuses, you will usually dissect the conceptus away from the uterus and perhaps the embryo away from the

> **Box 5.15** Visualization and Analysis of the Forming Skeleton
>
> The mature skeleton is composed of bone and permanent cartilages. Bone is formed by two different processes, endochondral and membranous ossification. Endochondral ossification is bone formation that occurs from mesenchyme condensations through a cartilage template. Membranous ossification is bone formation directly from a mesenchymal condensation. It is possible to show the developing cartilaginous skeleton starting at about 12.0 dpc by whole-mount Alcian Blue staining. Embryos are fixed whole without removal of the visceral organs, stained with Alcian Blue, and then cleared (MM3). Mineralization of the skeleton begins at later stages of development (~14.0 dpc). At these stages, one can stain for bone and mineralized tissues and also cartilage using a different method in which the fetus is skinned and gutted, fixed, and then processed for detection of bone and other mineralized tissues by Alizarin Red staining and cartilage by Alcian Blue staining (Figure 5.25). The tissues are digested in base and cleared in glycerol (MM3). The entire skeleton or specific regions of the skeleton can then be examined and documented.
>
> For photodocumentation, place the skeleton preparations in a glass petri dish in the clearing solution (for cartilage preparations) or plastic petri dish in glycerol (for Alizarin Red/Alcian Blue preparations). Image the skeleton preparations on a dissecting microscope with transillumination lighting. Image the mutant and control skeletons in the same orientation. You can take an image of the entire skeleton, usually a lateral view (Figure 5.25A,C), although it is also possible to take dorsal and ventral views. The limbs obscure the neck and thorax regions. However, if you are particularly interested in these regions, you can use forceps to remove the limbs. The limbs can also be imaged separate from the body, showing either dorsal or ventral views. The rib cage can also be imaged in isolation from the rest of the skeleton. Cut the ribs close to the vertebrae to release the rib cage from the rest of the skeleton. Then flat mount the dissected rib cage by placing a glass slide on top of it for subsequent imaging. The skull can be imaged as described in Chapter 6 (see Figure 6.4).

extraembryonic tissues (see Box 5.12). Use irrelevant pieces of the conceptus for genotyping, taking care not to contaminate extraembryonic tissues with maternal tissues. The choice of fixative to use will depend on the analysis you intend perform. For routine H&E histology, Bouin's fixative works very well. For RNA in situ hybridization studies, 4% paraformaldehyde (PFA) is best (MM3). For immunohistochemistry, 4% PFA may be adequate but you must determine which fixative preserves the relevant epitopes recognized by each particular antibody. For frozen sections, the fixed specimen is generally equilibrated into sucrose solution before freezing in OCT embedding medium in plastic molds. Orienting the specimen for histological sectioning is very important. Your mutant and control specimens must be consistently cut in the same plane to make useful comparisons to convince yourself and others (see Figures 5.13 and 5.18). Serial sections of the specimen are also very important to understand the three-dimensional morphology of the region in which the tissue is present. Cut sections of corresponding regions in mutants and controls. This may be difficult in the mutant because, by definition, the tissue or organ of interest is abnormal. Use unaffected regions of the mutant as reference points to compare with controls. When you generate images of the histological sections, it is a good idea to take images at low magnification to orient the viewer and also at high magnification to show specific details. Make sure that the histological staining or immunostaining is of consistent quality between mutants and controls so that one can focus on the genotype-specific, not technical, differences.

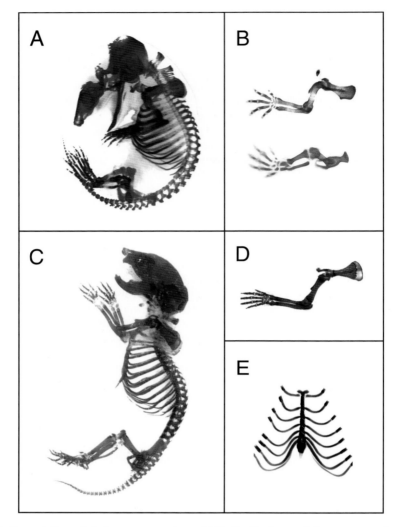

Figure 5.25. Presentation of skeleton preparations. (*A,B*) Cartilaginous skeleton preparation of 14.5-dpc embryo stained with Alcian Blue. Embryo is imaged in a glass dish in BABB (which dissolves plastic). Ammonium hydroxide treatment after fixation eliminates the yellow color caused by Bouin's fixative. (*A*) Intact embryo. If a dorsal or ventral image of the skeleton is desired, you can impale the lateral flank of the prepared embryo with a glass needle affixed to a support located to the left or right. (*B*) Forelimbs (*top*) and hindlimbs (*bottom*) were cut away from the skeleton preparation using a scalpel. (*C,D*) Skeleton preparation of newborn mouse stained with Alizarin Red and Alcian Blue. Skeleton preparation is imaged in a large plastic petri dish in 1:1 glycerol:ethanol. (*C*) Intact neonate skeleton. (*D*) Forelimb was removed using forceps to pull limb from prepared skeleton. (*E*) Rib cage was removed by cutting rib bones next to the vertebral column and a glass slide was placed on top of the rib cage to flatten it. Wild-type mice have seven ribs attached to the sternum (true ribs) and six ribs that are not attached to the sternum (false floating ribs). Not all of the false ribs are shown in this preparation. (*A–D* provided by Dmitry Ovchinnikov and Richard Behringer; *E* reprinted, with permission, from Rivera-Perez et al., *Development 121:* 3005–3012 [1995 ©Company of Biologists].)

> **Box 5.16** Quantitative Assessments of Tissue-specific Gene Expression
>
> One of the more difficult assessments to make in the molecular analysis of a tissue-specific defect that develops during embryogenesis is a quantitative one. Tissue-specific expression of molecular markers can be measured at the mRNA and/or protein levels. If you can isolate the specific tissue of interest, then you can use a variety of quantitative assays, including northern blot, semiquantitative or real-time RT-PCR, and western blot. However, the tissue of interest is often very small and not easily isolated. In these cases, expression is usually assayed by whole-mount or section in situ hybridization or immunohistochemistry. Can these in situ methods be used for quantitative assessments of gene expression? The answer is generally no, unless stringent controls are performed, and, even then, the assessment can only be semiquantitative. All samples must be processed using the same reagents, at the same time, for the same amount of time. This is especially important for non-radioactive detection methods that rely on color reactions that become more intense and saturated with time. If your gene of interest is also expressed in an unaffected tissue, you could potentially use that as an internal control for differences noted in the tissue of interest. Expression in the unaffected tissue should be the same between the mutants and controls.

5.E.2.b Molecular characterization of mutant phenotypes

As we discussed above (Section 5.D.2.b), molecular markers combined with knowledge of morphological defects and the expression pattern of the gene of interest provide essential information regarding the presence or absence of tissues or their state of differentiation. At later stages of embryogenesis, some methods of analysis become technically difficult. Whole-mount in situ hybridization methods must be modified to accommodate the larger size of the embryo and organs that hinders reagent access and enhances probe trapping. It is possible to bisect embryos, providing access of reagents to internal tissues. If you are studying one particular organ (e.g., the kidney, lung, heart, etc.), you can dissect it away from the embryo for whole-mount in situ hybridization. In situ hybridization of sectioned material circumvents these problems associated with whole-mount preparations, although three-dimensional information is harder to obtain. Remember that the lack of detectable expression of a molecular marker can be explained by two different mechanisms. First, the tissue may be present but not express the marker. Alternatively, the tissue may not be present and therefore there is no marker. To distinguish between these two possibilities, you need an independent assessment for the presence or absence of the tissue in question. In recent years, this has been facilitated because many null alleles have an associated reporter (e.g., *lacZ* or GFP) and the expression of the reporter can be used to identify the mutant tissue. Alternatively, either morphological features or the expression of an unrelated molecular marker may indicate the presence of a tissue. Molecular marker studies are, for the most part, qualitative assessments. Often one would like a quantitative assessment of gene expression within a particular tissue of a developing embryo. However, such assessments must be made very carefully (Box 5.16).

5.E.2.c Cellular characterization: Proliferation and death

Alterations in cell behaviors, including proliferation, death, and migration, can result in tissues that are absent, hypoplastic, or hyperplastic. Your morphological and histological analyses have

> **Box 5.17 Tissues That Can Be Cultured/Transplanted**
>
> **Tissues or cells that can be transplanted to ectopic or orthotopic sites**
>
> | | |
> |---|---|
> | hematopoietic tissues | pituitary |
> | liver cells | skin |
> | mammary glands | spermatogonial stem cells |
> | ovaries | testis tubules |
>
> **Organs that can be cultured transiently in vitro**
>
> | | |
> |---|---|
> | allantois | limb buds |
> | gonads | lung |
> | heart | mammary glands |
> | hematopoietic tissue | palate |
> | kidney | teeth |

been used to determine when the first tissue defects were observed. This time point serves as the starting point to examine cell proliferation and death (see Section 5.D.2.c). Alterations in cell proliferation or death may be transient rather than chronically present, and you may therefore have to perform your analyses at time intervals that are less than 24 hours to feel confident that you have not missed a transient alteration in these processes.

5.E.2.d In vitro analysis of cells and organs

Homozygous mutant embryos, whether or not the mutation is lethal, can provide a source of cells and organs for various types of in vitro analysis by culture. These embryos can be used to generate fibroblasts or other cell types for cell proliferation, cell death, and biochemical analyses. Traditionally, mouse embryonic fibroblasts (MEFs) are generated from 13.5-dpc embryos (MM3). However, it is possible to generate MEFs routinely from embryos as early as 8.5 dpc. Homozygous mutant and control cells can be used to examine cell proliferation by simple cell counts, cell cycle characteristics by fluorescence-activated cell sorting (FACS), and cell death by TUNEL, FACS analysis, or biochemical means. The chromosomes of these cells can be examined to study genome integrity. The homozygous mutant cells can also be used for biochemical studies or as a substrate for in vitro genetic manipulation to determine the position of the gene of interest in a biochemical pathway.

Homozygous mutant embryos may die before the desired stage to study an organ system of interest. However, for many organ systems, it is possible to dissect out the organ anlagen and culture it in vitro or in vivo for sufficient time to determine the consequence of the mutation in that organ (Box 5.17). For many in vitro organ culture systems, the organ of interest is cultured on top of a medium-permeable substrate (e.g., a filter) that sits above the medium (Figure 5.26). The organ grows at the air–medium interface. Some tissues or organs can be cultured submerged in medium. The cultured organs can then be analyzed for morphology, histology, molecular markers, cell proliferation, and death. In addition, the organs can be manipulated at

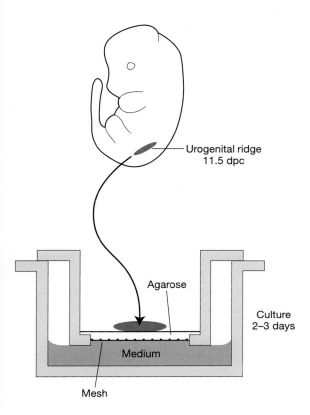

Figure 5.26. In vitro organ culture. In this example, the urogenital ridge containing the gonad and mesonephros is dissected from an 11.5-dpc embryo and placed in a cassette on a medium-permeable filter, in this case with a layer of agarose. This cassette is then placed into a well of medium. The organ is cultured at the air–medium interface for 2–3 days. During this period of time, the bipotential gonads will differentiate into morphologically distinguishable ovaries or testes, depending on their genotype. (Diagram courtesy of Akio Kobayashi and Richard Behringer.)

the initiation of the in vitro culture to monitor cell fates and migration. The use of fluorescent protein reporters can be used to dynamically image the development of the mutant organ in vitro.

Defects in cell migration can be another cause of tissue-specific morphology defects. Cell migration can be inferred by collating static images of different embryos that have been fixed at sequential stages of development. However, only correlative conclusions can be made from such a study. To make a strong conclusion about cell migration, one must perform more definitive studies. To document migration rigorously, cells must be indelibly marked either chemically or genetically and they and their cellular progeny observed over time in the same embryo or tissue. Lipophilic dyes such as DiI or DiO can be used to mark cells (by exposure or intracellular injection) of an embryo. However, you will be limited to specific times in development when the embryos and tissues are accessible for manipulation. It is relatively easy at embryonic stages that are compatible with postimplantation embryo culture and it is possible to mark cells in later-stage fetuses by exo utero manipulations, but these types of experiments are quite difficult. Alternatively, tissues or organs can be placed in organ culture for at least a day or two to follow marked cells. A drawback of chemical marking is that it becomes diluted as the marked cells proliferate.

A genetic strategy to mark cells exploits the DNA recombinase systems (e.g., Cre/*loxP* or Flp/*FRT*). For example, it is possible to generate mice that carry a Cre reporter allele that can be ubiquitously expressed (e.g., *Rosa26R*) and a CreER allele that is expressed in the tissue of interest (see Box 8.6, p. 193). CreER activity is induced by a pulse of tamoxifen, usually by intraperitoneal injection at a specific time in development when the cells of interest are thought to be migrating. At a later stage of development, the embryo is examined for the presence and distribution of reporter-expressing cells. One difficulty with this technique is that you cannot determine exactly which cells give rise to the marked cells. An alternative would be to use a Cre reporter allele that utilizes a fluorescent protein reporter. If the starting tissue can be imaged at the initiation of the induction of Cre activity and subsequently during culture, then you can follow the migration of marked cells in real time. Again, this will be restricted to those tissues and stages for which you have access.

Chimeras are useful tools for studying the actions of a mutation (Chapter 8). In a chimera study, the presence and distribution of mutant and wild-type cells are determined for a particular tissue by terminal analysis (e.g., β-gal staining and histology) of individual chimeras. The collation of the data from these chimeras provides insights into the actions of the mutation in that tissue. However, one of the technical conundrums of a chimera study is that individual chimeras or their tissues must be sacrificed at a single time point. Thus, one does not know the initial distribution of mutant and wild-type cells at the start of the experiment nor what the distribution of the mutant and wild-type cells would have been later. The development of fluorescent protein reporters can now be used to address this problem and obtain more information than previously possible from a chimera experiment. Accordingly, chimeras can be generated using mutant and wild-type embryonic cells. The mutant or wild-type cells can be marked with a ubiquitously expressed fluorescent reporter. At a specific stage of development, the organ of interest is explanted into in vitro culture and imaged to determine the presence, distribution, and behavior of the mutant or wild-type cells. The chimeric organ is then imaged during the culture period and the individual mutant or wild-type cells monitored. In this manner, the two genotypes of cells in the organs can be followed from start to finish.

CHAPTER 6

Phenotypic Analysis: Postnatal Effects

TWO PREVALENT MUTANT PHENOTYPES IN MICE are perinatal lethality and no overt mutant phenotype. These two types of phenotype may not be what most investigators desire after putting so much effort into generating a mutant mouse, but such are the complexities and realities of mammalian biology. Not to worry, because both types of mutant phenotype offer unique opportunities for understanding gene function.

You have arrived at this chapter because you have determined that homozygous mutants generated from heterozygous intercrosses are present among your mice at birth. In this chapter, we deal with postnatal mutant analysis. Of course, you may be fortunate and have a viable and fertile mutant with an obvious abnormality, thus making analysis much easier. But if not, we guide you in the analysis of perinatal and postnatal phenotypes, both viable and lethal. We also explain how to exploit a viable mutant that appears overtly normal—the dreaded "no phenotype" mouse. Many of the strategies and methods discussed in Chapter 5 can be used in the analysis of phenotypes detected postnatally. Of course, heterozygotes may have dominant postnatal effects that you would have initially identified in chimeras or in their progeny after germ-line transmission (Chapter 7).

Here, we start once again by examining the progeny of matings between heterozygous mice. Although this chapter offers a comprehensive plan for detecting abnormalities postnatally, you should, as always, let the expression pattern of the gene mutated be a guide for the possible phenotypes to expect and therefore the most useful investigations to pursue. If you have not previously investigated the postnatal and adult expression of your favorite gene, now is the time to do so in conjunction with the analysis of the mutant mice.

Chapter opening artwork courtesy of Andy Salinger and Monica Justice.

Are homozygous mutants born alive?

NO	Go to Section 6.A ▼
YES	Go to Section 6.B ▶

6.A Perinatal Lethality

You have determined that homozygous mutant pups generated from heterozygous intercrosses are present at birth, but they are found dead in the cage. You may have found intact dead pups or pieces of dead pups, depending on whether and how much the mother cannibalized the remains. DNA is relatively robust, so save the bodies or pieces for genotyping. First, determine the genotypes of all of the dead pups to find out whether the lethality is limited to the homozygous mutants or if heterozygous and wild-type pups are also among the dead.

There are two probable outcomes. In the first scenario, the dead pups are exclusively (or almost exclusively) homozygous mutants, suggesting that the lethality is limited to the homozygous mutant genotype. In the second scenario, homozygous mutants as well as heterozygotes and/or wild-type pups are among the dead. There are a couple of explanations for this outcome. It may have been a difficult birth, causing the mother distress and leading to the loss of not only the homozygous mutants but also, nonspecifically, the pups of the other possible genotypes. Thus, it is still possible that homozygosity for the mutation causes lethality. Examine additional litters to make this determination. The alternative and trivial explanation for the second scenario is that lethality is the result of poor husbandry or other environmental problems (noise, fumes, altered light cycle, temperature, diet, etc.) independent of genotype. Investigate and correct any husbandry/environmental problems and examine additional litters, coming back to the question at the beginning of this section: Are homozygous mutants born alive? If so, go to Section 6.B and if not, carry on.

Let us assume that you find that the dead pups are all, or nearly all, homozygous mutants. This is a good indication to move on to determine if the dead pups died before or soon after birth. You may have an indication of a catastrophic abnormality (e.g., top of the head open and no brain or herniated gut) that would very likely cause the death of the homozygous mutant pups. However, it is possible that the mother contributed to these abnormalities as she was cleaning the pups after birth. If the bodies are intact and not too degraded, you can do a very simple test to determine if the dead pups ever took a breath of air, assuming that they have a head, mouth, and lungs. Dissect the lungs and place them into a beaker of water. If the lungs sink, the mouse probably never took a breath after birth but might still have been born alive; if the lungs float, the mouse was alive at birth and took air into its lungs, but succumbed shortly thereafter. Do histology on the lungs to follow up on this initial conclusion, which might help determine the next steps in the analysis. If your mutated gene is expressed in the lungs, the lack of breathing could be a direct effect. On the other hand, there is a variety of other reasons why a pup may fail to breathe at birth (Section 6.A.1), including death before birth.

To pin down whether a mutant was alive or dead at the time of birth, one could attempt to be present at the time of parturition. However, this is basically futile because mice usually give birth during the dark phase of the light cycle, and the length of gestation is variable, give or take a day. The quickest and most direct way to determine if full-term fetuses are alive or dead just before birth is to set up timed matings (i.e., check plugs) and perform a Caesarian section (C-section) (MM3*) at 18.5 dpc. This is also very useful because you can bypass any abnormalities that might have been caused by the mother as she cleans the pups at birth by determining if catastrophic abnormalities observed previously are also present upon C-section. Wild-type pups should all be recovered alive following C-sections performed at 18.5 dpc and thus, if the homozygous mutants die late in gestation, you will detect them when you perform the C-section. These pups will not move, you will not be able to revive them, and they may appear necrotic. If the homozygous mutants are alive just before birth, you may see them moving and struggling to breathe following C-section, but they may never turn pink and will probably die within minutes after birth, indicating that perinatal lethality was caused by an inability to breathe. Alternatively, the homozygous mutant pups might move, breathe, become pink, and squeak, suggesting that they can survive birth (or at least a C-section), but then die shortly thereafter. A wild-type pup delivered by C-section should move vigorously. Observe the pups obtained from the C-section to determine if any are lethargic or in distress in any way. If the pups have all survived the C-section, foster them onto a surrogate female (see Box 3.4, p. 35) and observe them periodically to discover if any of the pups die or lag behind in their postnatal development. Genotype any pups that die to determine if they are the homozygous mutants. It is to be expected that the homozygous mutants will die soon after C-section if they were destined to die during natural parturition. However, it is possible that C-section could rescue a perinatal lethality, indicating that the phenotype is sensitive to the stress of labor and birth. This would provide a way of recovering live homozygous mutants to investigate the phenotype further. For each of the scenarios described above, analyze sufficient numbers of timed matings to do a statistical analysis to support your conclusions (see Box 5.1, pp. 72–73).

The homozygous mutants die before birth	Go to Section 5.E ▲
The homozygous mutants die at birth	Go to Section 6.A.1 ▼
The pups survive at least the first day after birth by C-section	Go to Section 6.B ▶

6.A.1 Analysis of perinatal lethality

There is a variety of reasons why a mutant mouse might die at birth. As mentioned above, catastrophic abnormalities are clearly incompatible with life after birth. If the mutants are intact and appear morphologically normal, then the most likely immediate cause of perinatal lethality is the inability to breathe or problems associated with the change in blood flow from the fetal to postnatal pattern that accompanies the first breaths. Mice can develop to term without a head or lungs, but they need to breathe to survive birth. Below we list some of the common causes of perinatal lethality, how to identify them, and how to study their development. Depending on the nature of the gene mutated and on its expression pattern, some of these causes may be more relevant than others in causing lethality of your particular mutant.

*MM3 refers to *Manipulating the mouse embryo: A laboratory manual*, 3rd edition (Nagy et al. [2003], Cold Spring Harbor Laboratory Press, Cold Spring Harbor, New York).

6.A.1.a Catastrophic abnormalities

A variety of catastrophic abnormalities can lead to the immediate death of a homozygous mutant newborn pup. If there is no head, brain, mouth, or lungs, then it will be impossible for the animal to breathe and it will die within minutes after birth. Another major defect that will lead to a quick death is herniation of the gut at the site of the umbilical attachment. If the guts are externalized, it is very possible that the mother may eat the externalized gut as she cleans the pup after birth. If the skin of homozygous mutant pups is not well developed, then the pups may die very quickly because of dehydration. These phenotypes develop before birth. Therefore, to understand the etiology of these mutant phenotypes, you will have to examine prenatal stages (see Chapter 5).

6.A.1.b Cardiovascular defects

With the first breath at the time of birth, dramatic changes occur within the heart and circulation concomitant with a shift in the site of oxygenation of the blood from the placenta to the lungs. The ductus arteriosus, which during fetal development shunts most of the blood leaving the heart through the pulmonary artery into the aorta, closes, and the pulmonary arteries dilate, allowing greater perfusion of the newly inflated lungs. At the same time, the increase in blood pressure in the left atrium due to the increased pulmonary circulation closes the foramen ovale, the interatrial opening that, during fetal life, allows oxygenated blood from the placenta to enter the left side of the heart to be pumped out through the aorta. The closing of these two shunts at birth separates pulmonary from systemic circulation, establishing the postnatal blood flow pattern (Figure 6.1). Failure of either of these shunts to close can cause respiratory and circulatory distress and result in neonatal lethality. Examination of the lungs, heart, and aortic arch arteries by gross and histological examination will reveal structural defects that can then be further explored during fetal life (see Section 5.E.1.b) to determine their developmental origin.

6.A.1.c Developmental delay

Developmental delay is a broad term indicating that a mouse is not fully mature when compared to age-matched controls. At birth, a pup with developmental delay may not have fully matured essential organ systems, such as the lungs, and would not be able to survive outside of the mother; it will likely be smaller than controls. Histological analysis of organs should be performed as a simple screen to assess the maturation of the various tissues in comparison to controls. Determination of the timing of the developmental delay will require timed matings and a prenatal analysis (see Chapter 5).

6.A.1.d Cranial nerve defects

Another reason why a newborn mouse may not survive the perinatal period is a defect in cranial nerve development. The cranial nerves comprise 12 pairs of nerves that derive from the neural crest and are essential for sensory and motor functions (Table 6.1). The simplest way to visualize the initial stages of cranial nerve development for comparison of homozygous

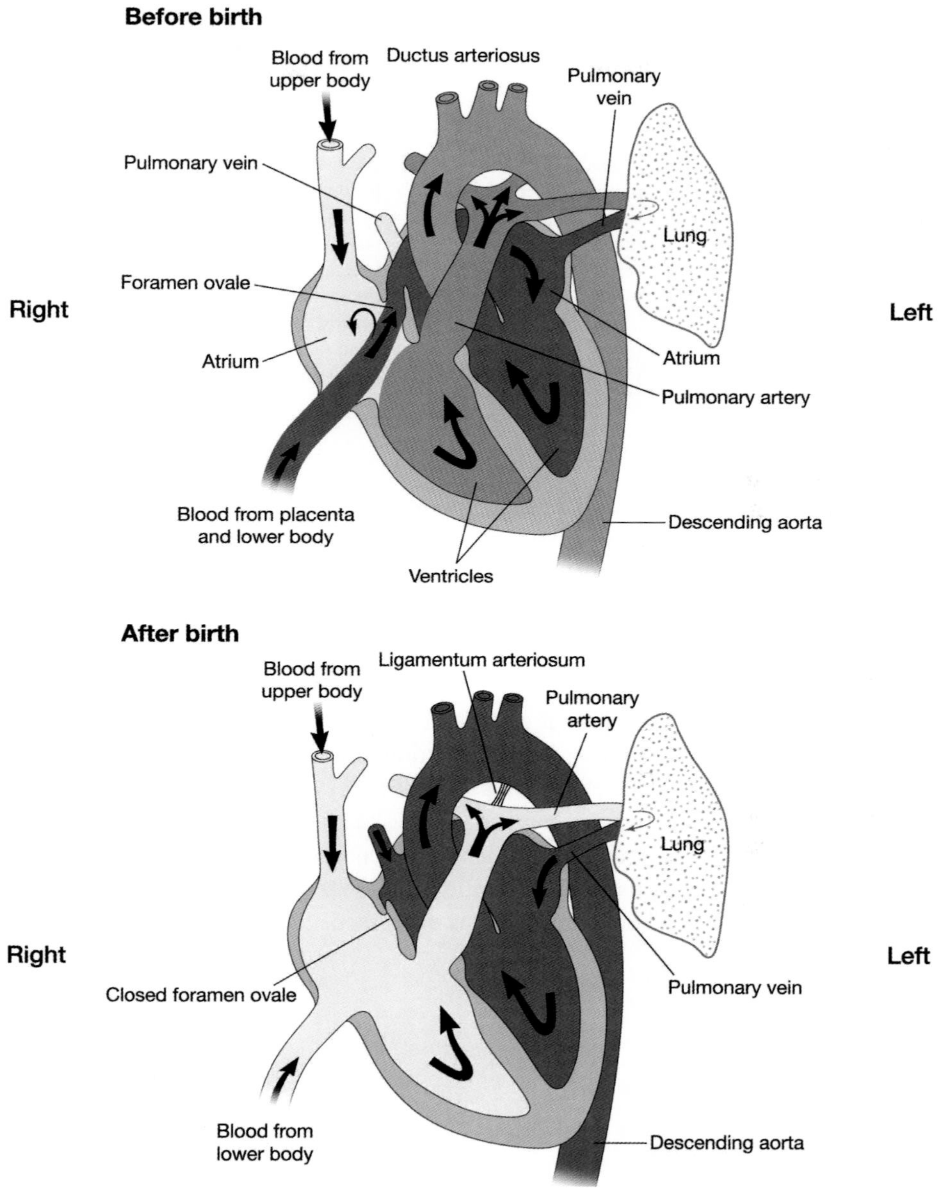

Figure 6.1. Diagram of cardiovascular changes at birth. During fetal life, oxygenated blood from the placenta enters the heart through the right atrium and is shunted through the interatrial septum via the foramen ovale to the left atrium. Blood is pumped into the ventricles through the right and left atrioventricular canals and is then pumped out through the aorta from the left ventricle and the pulmonary artery from the right ventricle. However, since the lungs are not functional and resistance in the pulmonary arteries is high, most of the blood leaving the left ventricle through the pulmonary trunk is shunted through the ductus arteriosus into the aorta. At birth, with the shutdown of the umbilical circulation, the initiation of breathing, and the inflation of the lungs, the ductus arteriosus closes, the pulmonary arteries dilate, and increased pressure in the left atrium closes the foramen ovale, thus separating pulmonary from systemic circulation. (Adapted, with permission, from *Langman's Medical Embryology*, 9th ed. [T.W. Sadler] Lippincott Williams & Wilkins Publishers [2004].)

Table 6.1. Cranial nerves

Number	Name	Modality	Function
I	olfactory	sensory	smell
II	optic	sensory	vision
III	oculomotor	motor	eye movement, pupil dilation
IV	trochlear	motor	eye movement
V	trigeminal	motor and sensory	head and face touch, pain; chewing
VI	abducens	motor	eye movement
VII	facial	motor and sensory	taste, ear touch, facial muscle movement
VIII	vestibulocochlear	sensory	hearing, balance
IX	glossopharyngeal	motor and sensory	taste; tongue, tonsil, pharynx touch; swallowing
X	vagus	motor and sensory	glands, digestion, heart rate
XI	spinal accessory	motor	head movement
XII	hypoglossal	motor	tongue movement

mutants and wild-type control littermates is to perform a whole-mount immunostain (MM3) at 10.5 dpc using the 2H3 antineurofilament monoclonal antibody available from the Developmental Studies Hybridoma Bank, Iowa (Figure 6.2).

6.A.1.e Skeletal defects

Abnormalities in skeleton development can also cause perinatal lethality. For example, if the rib cage is too small or split, the mouse will have problems breathing (Figure 6.3). It is probably a good idea to generate a skeleton preparation (MM3) of any perinatal lethal mutant. The skeleton is a robust tissue. Thus, you can generate a skeleton preparation even for pups found dead in the cage. The skin or viscera can be used for genotyping. Compare the skeletons of the mutants with controls, paying particular attention to the size and structure of the rib cage and any other areas in which you know your gene is expressed.

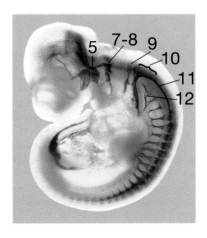

Figure 6.2. Whole-mount immunostaining to see the cranial nerves. Wild-type, 10.5-dpc embryo immunostained using an antineurofilament antibody. Numbers indicate cranial nerves. (Image courtesy of Deborah L. Guris and Akira Imamoto.)

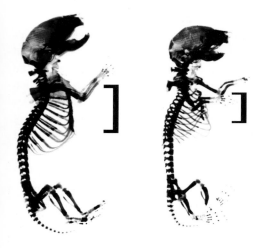

Figure 6.3. Skeleton preparation to visualize cartilage and bone. Wild-type (*left*) and $Sox9^{+/-}$ (*right*) newborn mice. Brackets highlight the difference in size of the rib cages between the mutant and control. (Image courtesy of Benoit de Crombrugghe.)

6.A.1.f Cleft palate

Cleft palate is a relatively common phenotype of mutant mice that causes death soon after birth. The mice may be born alive and breathe air, but much of the air ends up in the stomach, causing a bloated belly. To determine if the mutant neonate has cleft palate, it is best to remove the head from the body, remove the jaw, and view the upper part of the head from the ventral side to see the palate. If there is a cleft of the soft palate, it will be morphologically obvious (Figure 6.4A). If you identify cleft palate, the next step is to generate a skeleton preparation (MM3) to view the skull bones (Figure 6.4B,C). The maxillary and palatine shelves form the

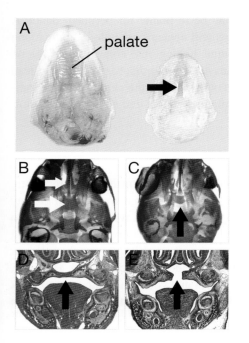

Figure 6.4. Morphological and histological analysis of cleft palate. (*A*) Ventral view of the upper portions of the heads of wild-type and *Dph2P1* homozygous mutant newborn mice, showing the roof of the mouth. The palate region of the control (*left*) is intact, whereas the mutant (*right*) has cleft palate (*arrow*). Note that in this case, the age-matched mutant is also smaller because of developmental delay. (Image provided by Chun-Ming Chen and Richard Behringer.) (*B,C*) Skeleton preparations showing ventral views of the skulls of wild-type (*B*) and *Tbx1* homozygous mutant (*C*) newborn mice. The jaws have been removed to more easily view the bones of the palate. (*Large white arrow*) Palatine shelf; (*small white arrow*) maxillary shelf; (*black arrow*) absence of bone regions. (*D,E*) Histology of palate formation in wild-type (*D*) and *Tbx1* homozygous mutant (*E*) 16.5-dpc embryos. Frontal sections show the presence and absence of palate closure (*arrows*). (*B,C* Adapted, with permission, from Jerome and Papaioannou, *Nat. Genet. 271:* 286–291 [2001 ©Nature Publishing Group]; *D,E* courtesy of Robert Kelly, Loydie Jerome-Majewska, and Virginia Papaioannou.)

hard palate of the skull. An easy way to observe these bones in a skeleton preparation is to remove the skull from the body and remove the lower jaw from the skull. You can then take pictures of the ventral side of the skull to examine the palate. The palate forms around 14.5 and 15.5 dpc when the bilateral palatine shelves fuse along the midline. Therefore, analyze the mutants by histology between 11.5 dpc, as palatal shelf formation initiates, and 15.5 dpc, when they fuse. Frontal sections of the head provide the best view of the forming palatal shelves (Figure 6.4D,E). A three-dimensional view of the forming palate can be obtained with scanning electron microscopy (SEM). Depending on what you observe in your histological and/or SEM analysis and on the nature and expression pattern of the gene, you may want to consider measuring cell proliferation and apoptosis during palate formation as well as analyzing relevant differentiation markers by in situ hybridization or immunohistochemistry.

6.A.1.g Diaphragm defects

Another cause of breathing problems is a diaphragm defect. A hernia of the diaphragm can be detected grossly on careful dissection. You may find organs of the peritoneal cavity pushed up through the hernia into the chest cavity. Alternatively, the diaphragm may be too thin to support breathing. This is best assessed by histology in comparison to controls.

6.A.1.h Other causes of perinatal lethality

A defect in the breathing centers of the brain could also lead to perinatal lethality. This is difficult to determine, but one could examine the neural circuitry that regulates breathing by histology or marker gene analysis, particularly in areas of expression of the gene. This would at least provide correlative evidence for a central nervous system (CNS) defect.

Additionally, other defects unrelated to breathing may be to blame for perinatal death. A severe anemia could lead to perinatal death. Anemic pups will appear very pale in comparison to controls. An analysis of the blood (red blood cell counts, hemoglobin concentration, blood smears, etc.) would be a reasonable line of investigation. A bleeding disorder could also lead to perinatal lethality. Blood in the gut or abdomen, or blood spots on the skin, would be indications of a bleeding disorder. In these cases, measure platelet counts and analyze blood clotting factors.

6.B Analysis of Postnatal Mutant Phenotypes

We define postnatal mutant phenotypes as those observed on the day of birth but not including those that cause perinatal lethality. Thus, you are presented with pups that are alive on the day of birth. The next step is to observe their growth and development to weaning and on into adulthood. If the pups survive 7–10 days, you will be able to take a tail sample for genotyping and mark them by toe clips. Likewise if they survive to weaning, you can number them by ear punching (MM3). These are indelible marks that allow you to identify each animal. However, if the homozygous mutants survive birth but die before 7 days and you would like to keep track of the homozygous mutant from the day of birth, you will need an alternative marking system (Box 6.1).

One of the first steps to take with your homozygous mutant is to examine it for gross abnormalities, focusing on known areas of gene expression. This examination can be carried out at

> **Box 6.1** **Marking Newborn Pups**
>
> Many times you may wish to genotype the newborn pups from a heterozygous intercross so that you can follow the development of a mutant phenotype starting at birth. It is easy to cut a small piece of tail from the newborn pups for polymerase chain reaction (PCR) genotyping, but how do you mark the pups to match them with their genotypes? One crude but effective method is to use a marking pen to number each pup, but you will have to check periodically to touch up the mark if it begins to fade. Once the pups are old enough, use a traditional marking system (toe clip or ear punch) to keep track of the mice. Another strategy is to clip the toes to mark the neonates. However, the toes of mouse pups have not yet separated at birth, making toe clipping difficult. It is possible to use fine scissors to cut the most distal part of the toe to create an indelible mark. The small drops of blood from the toe and tail cuts do not usually seem to bother the mothers. **Before performing any of these procedures, check with your local animal care committee for appropriate approvals.**

any time between birth and adulthood, depending on the characteristics of your gene and mutation, e.g., when it dies if it is lethal. Examine the external features of both males and females using the checklist provided in Figure 6.5 as a guide. Next, sacrifice the homozygous mutants, both males and females, and perform an internal examination of the visceral organs. Use the

- ☐ Body size: males bigger than females of same age
- ☐ Teeth: normal occlusion, incisors, and molars present
- ☐ Whiskers
- ☐ Eyes normal size and open by 14 days
- ☐ Head shape
- ☐ Ears
- ☐ Toes on forelimbs and hindlimbs (five on each); toenails on the dorsal side of each digit
- ☐ Hair expected color and normal texture
- ☐ Genitalia of male or female
- ☐ Tail length and straightness

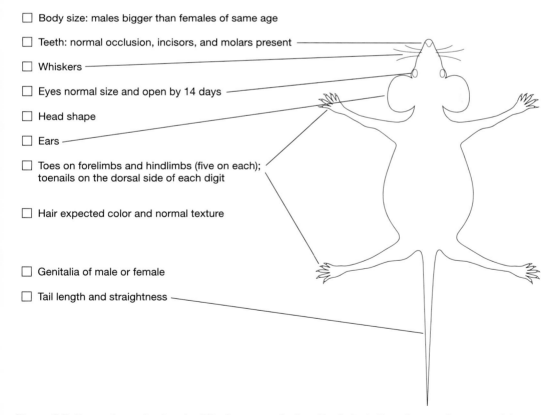

Figure 6.5. External examination checklist for postnatal mice. Check the indicated external organs and tissues to determine if they are present and grossly normal relative to wild-type age- and sex-matched littermates.

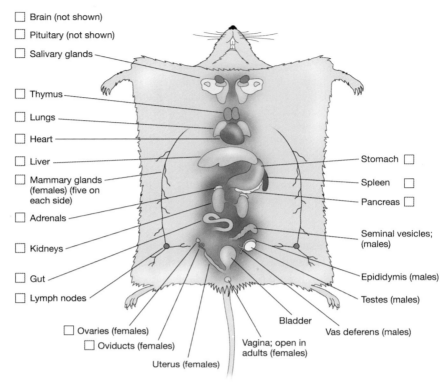

Figure 6.6. Internal examination checklist for postnatal mice. Check the indicated internal organs and tissues to determine if they are present and grossly normal relative to wild-type age- and sex-matched littermates. (Adapted from *Biology of the Laboratory Mouse*, 2nd edition. Roscoe B. Jackson Memorial Laboratory [ed. E.L. Green et al.], Dover Publications [1975], courtesy of The Jackson Laboratory.)

checklist provided in Figure 6.6 as a guide. The male and female reproductive organs are shown in Figure 6.7. A guide to the distribution of the five pairs of mammary glands is shown in Figure 6.8. After performing this analysis, you can answer the following question:

Do the mutants have a visible abnormality (one that includes an obvious external and/or internal defect[s])?

YES	Go to Section 6.B.1 ▼
NO	Go to Section 6.B.2 ▶

6.B.1 Visible mutant phenotypes

Weaning is the transition a pup makes from feeding on milk from its mother to eating solid food on its own. Normally, this will occur when you remove the litter from the mother, but even if the offspring are left with the mother, they will begin to depend more and more on solid food

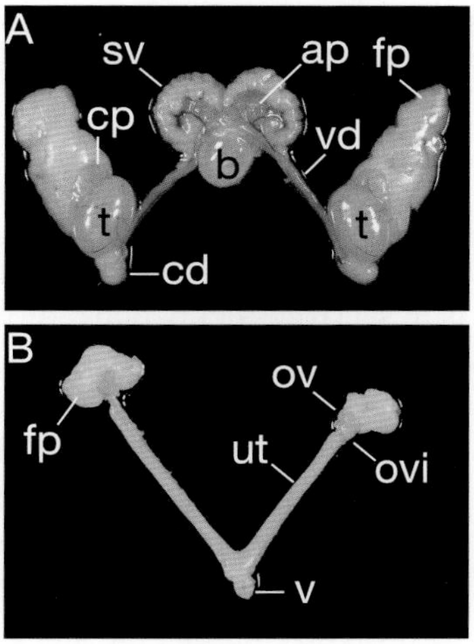

Figure 6.7. Male and female reproductive organs. Male (*A*) and female (*B*) reproductive tracts. (ap) Anterior prostate; (b) bladder; (cd) cauda epididymis; (cp) caput epididymis; (fp) fat pad; (ov) ovary; (ovi) oviduct; (sv) seminal vesicle; (t) testis; (ut) uterine horn; (v) vagina; (vd) vas deferens. (Reprinted from Jamin et al. [*Nat. Genet. 32:* 408–410 [2002 ©Nature Publishing Group].)

starting at about 3 weeks postnatally. Thus, weaning is an important milestone in mouse development. Many aspects of life before weaning can be provided by maternal care and feeding. Once the pups are weaned, they are completely on their own and are thus challenged to survive. Accordingly, we have split the analysis of postnatal visible mutant phenotypes into before weaning and after weaning, and also we consider age-related phenotypes that appear in relative old age. Many phenotypes, such as morphological abnormalities, are evident throughout all of these periods, but some only appear after weaning, at the time of sexual maturity, or during old age.

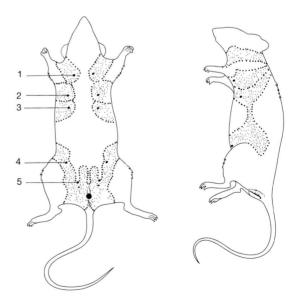

Figure 6.8. Mammary glands in the mouse. Ventral (*left*) and lateral (*right*) views showing the positions of the five pairs of nipples and mammary glands located subcutaneously, attached to the dermal side of the skin. (Adapted from *Biology of the Laboratory Mouse*, 2nd edition, Roscoe B. Jackson Memorial Laboratory [ed. E.L. Green et al.], Dover Publications [1976], courtesy of The Jackson Laboratory.)

6.B.1.a Visible mutant phenotypes before weaning

Preweaning lethality. Lethality will usually be presaged by the appearance of a sickly pup or one that is not thriving like its littermates. However, because you cannot watch your mice continuously, you may also simply find them dead in the cage. There is always a temptation to let a sickly pup continue to live but more often than not, it will die when you are not present and either be cannibalized or degrade, precluding some types of analysis. The best course of action is to sacrifice sickly pups for genotyping and phenotypic analysis as soon as they are detected.

One simple reason that a mouse might die, even if it initially appears active and vigorous, is that it does not or cannot feed. Complete lack of feeding would cause death very quickly, not only because of a lack of nourishment but also because of the lack of liquid normally obtained from the milk. You can easily determine if a pup is feeding during the first week postnatally by looking at the left side of the abdomen. A mouse that has been nursing will have its stomach full of milk that is white and this can be seen through the translucent skin (Figure 6.9). By the way, if you find that the stomach is on the right side of the abdomen, then you have identified a defect in left-right patterning. Other reasons for no milk in the stomach could be cranial nerve defects (see Section 6.A.1.d and Figure 6.2) or the inability to emit sounds. Mouse pups emit ultrasonic vocalizations that elicit maternal behavior such as retrieval of scattered pups to the nest. In the absence of vocalizations, the homozygous mutant pups are more likely to be left out of the nest and fail to nurse. Incidentally, if all of the pups, mutant and wild type, are scattered outside the nest, it could indicate a dominant hearing or olfaction problem on the part of the heterozygous mother because both of these senses are important for maternal recognition of the young.

Sudden death during the preweaning (or even postweaning) period that is not accompanied by eating problems could be attributable to defects in the conduction system of the heart that result in spontaneous ventricular arrhythmia or tachycardia. Conduction system defects may or may not be accompanied by structural defects, so functional tests such as an electrocardiogram (EKG) or a test that assesses heart rate may be called for.

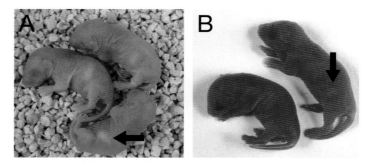

Figure 6.9. Detection of milk in the stomach of newborn mice. (*A*) Three neonates are shown. A white milk spot (*arrow*) is seen on the left side of their abdomens through their translucent skin. (*B*) Wild-type (*right*) and elastase-diptheria toxin transgenic (*left*) mice. These mice have not yet fed. In the wild-type mouse, a triangular white patch (*arrow*) is seen through the skin on the left side of the pup. This is the pancreas, not milk in the stomach. Elastase-directed expression of the toxin in the pancreas causes ablation and hence no white patch.

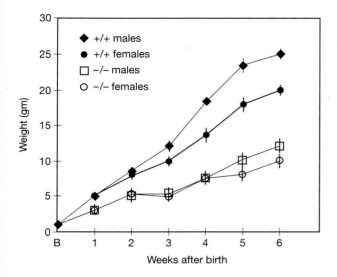

Figure 6.10. Growth curves of wild-type and growth-deficient mice. Pups generated from *Otx3* heterozygous intercrosses were marked at birth and weighed at 1-week intervals. The weights of wild-type males and females diverge between 1 and 2 weeks after birth. This sexual dimorphism is maintained as the mice age. The rate of growth of *Otx3* homozygous mutant males and females is always less than that of the controls. (Image courtesy of Aki Ohtoshi and Richard Behringer.)

Altered growth pattern. One prevalent phenotype in mutant mice is a growth defect. Affected mice appear otherwise healthy but are smaller than controls. You can document a growth defect by generating a growth curve for the mutants and controls. Male and female mice are sexually dimorphic with respect to body size. Therefore, you will need to segregate your data by sex. Generate a sufficient number (>5) of each genotype and sex to make a statistically significant judgment on any growth differences. Weigh each mouse at regular intervals (e.g., daily, every other day, or weekly) and generate a growth curve. Be sure to include error bars for each genotype at each time point to indicate the observed variation. An example of growth curves from wild-type and mutant mice with a growth defect is shown in Figure 6.10. From birth to weaning, mice gain weight rapidly. There may be a small dip in the rate of growth around weaning as the pups make the transition from milk to solid food. At about 6 weeks of age, the rate of growth slows and eventually plateaus. A mutant with a growth deficiency will have a growth curve that diverges significantly from wild type.

A fundamental question to address is whether the growth deficiency manifests itself before or after birth. Accordingly, determine the weight of the mutants relative to controls obtained from a C-section at 18.5 dpc. A C-section will bypass variation that might occur on the day of birth if the mice have fed and have varying amounts of milk in their stomachs. If you find that the mutants and controls are of similar weight at 18.5 dpc, then the growth deficiency developed postnatally. The neural regulation and pituitary control of somatic growth would be one avenue of subsequent investigation. Alternatively, a growth deficiency could be caused by a skeletal defect, metabolic defects, hyperactivity, or dental problems.

Morphological abnormalities. If your examination of the external and internal features of the homozygous mutants (Figures 6.5 and 6.6) has identified an overt defect in a tissue or organ system, the tissue or organ may be absent, hypoplastic, or abnormal in morphology. You may have already anticipated such a phenotype because of the expression pattern of the gene mutated. First, document the gross abnormality in the intact animal, either awake or anesthetized, if the morphological abnormality is externally visible. If the defect is internal, sacrifice the mouse and perform a dissection to reveal the affected tissue or organ and take a picture. It is always important to take pictures with a size marker (e.g., a ruler). If the mutant organ is present, dissect it out and photograph the isolated organ. Remember to photograph controls also. The next step will depend on the type of tissue or organ that is affected. If you find a skeletal defect, then a skeleton preparation (MM3) would be informative. Generally, though, the next step would be to perform a histological analysis of the affected tissue or organ to understand which cell types are altered. Subsequently, gene expression studies, including in situ hybridization and immunohistochemistry of molecular markers, should provide insights into the molecular defects associated with the abnormal morphology. Cell proliferation or death assays will provide further insights into tissue defects. Finally, you will need to understand the development of the morphological abnormality and perform a prenatal analysis to determine the primary defect and its time of onset (Chapter 5).

Neurological problems. According to the Mouse Genome Informatics (MGI) Phenotype Ontology Browser, behavior/neurological phenotypes are classified as the following:

- abnormal feeding/drinking behavior
- abnormal learning/memory/conditioning
- abnormal motor capabilities/coordination/movement
- abnormal sensory capabilities/reflexes/nociception
- abnormal sleep pattern/circadian rhythm
- abnormal social/conspecific interaction
- neurological/behavioral: no defect detected
- other abnormal behavior
- seizures

These classifications provide standardized phenotype categories to consider as you assess your mutants for potential neurological disorders. Some of these phenotypes are more obvious than others to detect. Eating-behavior defects may become immediately obvious if such homozygous mutant pups are smaller than their littermate controls. However, overeating may not be immediately obvious because it takes longer to accumulate weight. Learning and memory alterations as well as sleep pattern/circadian rhythm defects are not obvious and require sophisticated tests using specialized equipment for assessment. If you suspect these types of phenotypes, then it is probably best to enlist a collaborator with the appropriate expertise. Abnormal social behavior, especially aggressive behavior, may become apparent because you will notice that heterozygous and wild-type mice in cages with homozygous mutant mice develop injuries (e.g., scars on their backs). Seizures, indicative of CNS defects, will also be obvious but only if you are present when they occur.

Below we discuss the analysis of a subset of neurological phenotypes. Abnormalities in motor capabilities, coordination, or movement are among the most obvious neurological phenotypes. Thus, if you have a mutant phenotype of this class it will likely be very obvious. Abnormalities in sensory capabilities, reflexes, or nociception are not overt phenotypes but can be assessed using a set of simple methods that are covered in Section 6.B.2.a.

Abnormalities in motor capabilities/coordination/movement. At birth, mouse pups can crawl sideways. At 3–4 days after birth, they can pivot. By approximately 6–7 days, they can walk and at 9 days, they can run unsteadily. By 15 days postpartum, the pups can move like adults. Table 6.2 lists the phenotype ontology of abnormal locomotor activity. These are precise standardized definitions of locomotor behavioral defects.

There are many causes of an abnormal locomotor phenotype. The first thing to do is to document when the mutant behavior is first detected and whether it becomes progressively stronger or milder as the mice age. This will give you an indication of when the cellular, tis-

Table 6.2. Abnormal motor capabilities/coordination/movement subcategories

Akinesia	absence of movement or loss of the ability to move, such as temporary or prolonged paralysis or "freezing in place"
Ataxia	inability to coordinate voluntary muscular movements
Bradykinesia	decreased spontaneity and movement
Dystonia	state of abnormal tonicity (hypertonicity or hypotonicity) in any tissue resulting in impairment of voluntary movement
Hyperactivity	general restlessness or excessive movement; more frequent movement from one place to another
Hypoactivity	reduced movement from one place to another
Impaired coordination	reduced ability to execute integrated movements of muscle
Increased vertical activity	greater than average time spent jumping or rearing
Jumpy	marked by fitful, jerky movements
Lethargy	mild impairment of consciousness resulting in reduced alertness and awareness
Negative geotaxis	mice, when placed on a downward slanting grid, walk down without turning around, whereas wild-type mice will always turn around and walk upward
No spontaneous movement	failure to make any change in position or posture
Paralysis	loss of power of voluntary movement in a muscle through injury or disease of it or its nerve supply
Peculiar gait	unusual or distinctive way of walking
Reduced vertical activity	less than average time spent jumping or rearing
Retropulsion	when placed in a new environment, mice will walk backward and then may walk forward, whereas wild-type mice will immediately walk forward, or freeze momentarily and then walk forward
Short stride length	reduced average distance between steps
Stereotypic behavior: backflips, circling, excessive scratching, head bobbing, head shaking, head tossing, increased stereotypic behavior, jerky movement, spinning	repetitive, invariant, perseverative motor patterns that do not appear to be purposeful
Walking backward	locomotor activity in the posterior direction
Wild running	self-explanatory

> **Box 6.2** Video Documentation of Mice
>
> Set the stage so that the viewing area does not have anything in it that will distract the viewer from watching the behavior of the mouse. For example, have a platform or cage with only the test mouse. If you use a platform, cover it with a sheet of white filter paper. If you use a cage, use one with fresh bedding. You will quickly realize that mice urinate and defecate as a stress response to a new environment. Simply wait a few minutes until they have nothing else to produce, clean up the platform or cage, and then begin imaging. Use a white or light-colored backdrop if you are shooting the video from a lateral view, unless the mouse is albino, in which case a darker background might be better.

sue, or organ defect first becomes detectable and the time frame of any changes. Document a neurological phenotype on video (Box 6.2) and remember to also document wild-type controls for comparison. Document sufficient numbers of homozygous mutants and controls to come up with a consistent assessment of the mutant phenotype.

Myelination, the formation of myelin on axons, insulates axons for fine motor control and also functions to speed nerve conduction. Thus, defects in myelin can lead to abnormalities in motor control and nerve conduction. In the mouse, myelination of the neurons that regulate motor control occurs at approximately 10–14 days postpartum. Thus, locomotor defects caused by myelin dysfunction will probably not manifest themselves until about 2 weeks after birth. Ataxia or "shivering" phenotypes that become apparent at this stage suggest a myelination defect. Developmental defects in locomotor coordination could also suggest an abnormality in the cerebellum. This region of the brain functions to regulate movement and balance. In the mouse, a significant amount of cerebellar development and maturation occurs after birth. The major neuronal cell types in the cerebellum (granule and Purkinje cells) undergo migration, axon extension, dendrite formation, and synaptogenesis until the end of the third week after birth. Thus, abnormalities in movement and balance that become apparent after 14 days postpartum may indicate a cerebellar defect.

Investigate whether the neurological phenotype is more likely caused by a defect of the CNS, peripheral nervous system (PNS), or the vestibular system. Again, let the expression pattern of the gene in question serve as a guide for which tissues to analyze. The first step would be to do a histological analysis of the brain and peripheral nerves (e.g., the sciatic nerve). Standard hematoxylin & eosin (H&E) staining is sufficient for a superficial look at the brain. Stains that are useful for detecting defects in myelin include Luxol Fast Blue or Gold Chloride. You can also do immunostaining for myelin basic protein (MBP) or myelin proteolipid protein (PLP) for the CNS, and MBP or peripheral myelin protein zero for the PNS. With a decrease in myelin, you will see normal numbers of axon bundles but a reduction in myelin around the bundles. If there is a neuronal defect, you may see fewer axons but the axons that are present will have normal amounts of myelin. Combinations of these two phenotypes are also possible. It is also possible that the histology of the neural tissues will be uninformative because of a defect in a discrete set of neurons. Knowing the precise expression pattern of the gene of interest will be very important in this situation. However, if the gene is ubiquitously expressed, then the analysis will be harder. If this is the case, you will need to perform more detailed neurological tests to learn more about the phenotype.

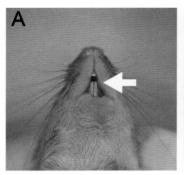 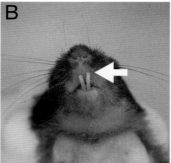

Figure 6.11. Mouse incisors. The top and bottom pair of incisors of a mouse grow continuously. Because they are aligned with one another and meet, they are constantly maintained at a certain length (*A*). If the incisors are misaligned (*B*), they will not meet and will continue to grow, causing the mouse difficulties in eating. Such teeth must be trimmed so that the mouse can feed. (Images courtesy of Kin Ming Kwan and Richard Behringer.)

6.B.1.b Visible mutant phenotypes after weaning and sexual maturity

Compromised growth and vigor. Most abnormalities will be picked up before weaning with a thorough internal and external exam (see Figures 6.5 and 6.6) and careful observation of the mutant mice. However, a growth defect or failure to thrive that only becomes evident after weaning may be due to problems with eating or digestion of solid food. One reason that a pup may not thrive around the time of weaning is an absence of teeth or malocclusion of the incisors. Without teeth or with abnormal teeth, the pups will not be able to eat the hard food pellets that are the usual diet of laboratory mice. To circumvent this problem, mice without teeth can be fed a commercially available soft mouse chow. Alternatively, you can make your own soft mouse food (Box 6.3). The incisors of mice grow continuously, but because the upper and lower incisors meet and are worn down by gnawing, they are maintained at a constant length. If the incisors are misaligned, then they will become excessively long, preventing the mouse from eating properly (Figure 6.11). If this is a feature of your mutant phenotype and you want the mice to survive, you will have to trim the teeth with strong scissors and perhaps put them on a soft food diet.

Box 6.3   **Recipe for Soft Mouse Food** (Courtesy of Phil Soriano)

Ingredients
    10 g agar
    100 g commercial (human) baby formula
    300 g powdered mouse chow (grind it up or scrape the bottom of the barrel)

Dissolve the agar in 300 ml of water in a microwave oven, on high.
Dissolve the baby formula in 500 ml of water at room temperature.
Combine the melted agar and baby formula, stirring vigorously.
Quickly add powdered mouse chow while stirring.
Chill and serve.

Compromised growth and vigor could also be indications of problems with digestion and metabolism. If the teeth of the homozygous mutant appear normal, then perhaps there is a defect in the gastrointestinal tract (gut, pancreas, etc.). Another type of gastrointestinal defect that can cause abnormal growth and even death is called megacolon. Mice with megacolon have defects in their ability to defecate. Alternatively, there may be abnormalities in the animals' metabolism. Clinical chemistry assays can be used to pinpoint the accumulation or deficiency of a metabolite, potentially guiding you to the primary defects in a metabolic pathway.

Infertility in males. Between 4 and 6 weeks, another previously cryptic phenotype, infertility, may become apparent. If your mutated gene is expressed in the gonads during development or in the adult, this is an obvious phenotype to look for. However, many factors contribute to successful reproduction so that gonadal gene expression is not a prerequisite for effects on fertility. Here we present an overview of how to identify and analyze a male fertility defect. At some point if the analysis progresses more deeply, you may want to seek the advice of an expert on male reproduction. As a starting point, you can simply breed the males to determine if they are fertile. However, if you have generated extra homozygous mutant males and you already suspect a potential effect on spermatogenesis, you can do a few simple tests without breeding that can potentially get you to a phenotype more quickly. First, determine if you have recovered the predicted proportion of homozygous mutant males from heterozygous intercross matings. Do a statistical analysis to reach a conclusion (see Box 5.1, pp. 72–73). If there are too many males, determine the genetic sex of all of the homozygous mutants (see Box 3.7, p. 40). This will indicate whether you have phenotypic males that are genetically female (XX). XX males are invariably sterile because they lack the Y chromosome that carries essential spermatogenesis genes and they have too many X chromosomes that interfere with spermatogenesis. You should have already determined that the homozygous mutant males look like males and have all of the male-specific organs (see Figures 6.5–6.7).

Next, check the cauda epididymis. The cauda epididymis of an adult male should be white in appearance because of the mature sperm that it holds. A cauda epididymis that is small and translucent is an indication of a sperm defect. Release the sperm from the cauda epididymis as if you were performing an in vitro fertilization experiment (MM3). If there is sperm, is it motile? If there is no sperm or the sperm is immotile, you may have revealed a potential fertility phenotype. Check sufficient numbers of homozygous mutant males and sibling controls to determine if the phenotypes that you are observing are only observed in the homozygous mutants.

Although many fertility mutants can be identified using these methods, there are others that will appear to be wild type by those criteria. Thus, ultimately, you will have to perform a breeding test. Breed homozygous mutant males with wild-type females and wait for a pregnancy. If you set up a timed mating, you will be able to determine if the homozygous mutant males can produce a plug. Females from timed matings can then be sacrificed during gestation to determine if they are pregnant. If you find embryos, then of course the male is fertile. To be thorough, it is probably a good idea to genotype the embryos to verify paternity. They should all be heterozygotes. This still does not verify that the males are completely wild type with respect to spermatogenesis because as little as 10% sperm production is sufficient for fertility.

If no pregnancies are obtained from the homozygous mutant males, then the next step is to determine if their sperm can fertilize oocytes. Set up timed matings with wild-type females to obtain plugs. Collect the oocytes from the oviducts (MM3) and view them under the dissecting microscope to determine if they have been fertilized (i.e., are two pronuclei present? See Figure 5.4B). If they are not fertilized, then the sperm from the homozygous mutant are likely to be defective. If there are two pronuclei, then culture the embryos and determine the percentage of fertilized oocytes that progress to the blastocyst stage (see Box 5.4, p. 82). Compare these numbers with wild-type controls. It is possible that the mutant sperm can fertilize the oocytes but then development fails. This would also suggest that the mutant sperm are defective.

At this point, you know if your homozygous mutant males can produce a plug, whether they have motile sperm, and whether the mutant sperm can fertilize oocytes. What do you do next? Spermatogenesis is a multistage process that is well characterized. If no sperm are present in your mutant, it is likely that they lack germ cells or have a block in spermatogenesis. On the other hand, the production of defective sperm that is immotile or cannot fertilize oocytes is likely caused by abnormalities in spermiogenesis, the transformation of a round haploid cell called the spermatid into the morphologically distinct spermatozoa.

At birth, the testes only have spermatogonia, the stem cells that produce the spermatozoa. By 10 days after birth, a cohort of the spermatogonia has progressed into meiosis to generate spermatocytes. By 19 days, they have become haploid, round spermatids. The first mature spermatozoa are subsequently formed at approximately 30 days after birth. This initial formation of sperm after birth is called the first wave of spermatogenesis (Figure 6.12). Subsequent overlapping waves generate the variety of stages present in the adult testes. In adult males, it

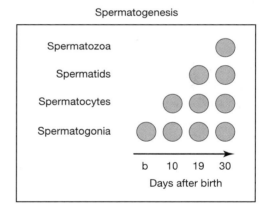

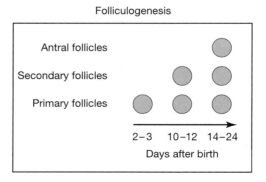

Figure 6.12. The first waves of spermatogenesis and folliculogenesis. (*Top*) Diagram showing the various types of sperm cells present at different days after birth (b), during the first wave of spermatogenesis. At birth, only spermatogonia are present in the testes. With time, more differentiated sperm cell types are present in the postnatal testis. Meiotic spermatocytes are present from day 10 onward and haploid spermatids are present from day 19 onward. The first mature spermatozoa are present at 30 days after birth. Subsequently, the length of time needed to generate spermatozoa increases to about 41 days. (*Bottom*) Diagram showing the various types of follicles present at different days after birth during the first wave of folliculogenesis. By 2–3 days after birth, all oocytes are in meiotic prophase in primary follicles. With time, more differentiated follicles are present in the postnatal ovary. Secondary follicles are present by 10–12 days after birth and antral follicles are present by 14–24 days after birth.

takes approximately 41 days for spermatogonial stem cells to form mature spermatozoa. Thus, you can histologically analyze homozygous mutant males at various times after birth to determine when spermatogenesis goes wrong. Alternatively, you can analyze adult testes histologically to determine which stages of spermatogenesis are abnormal. Transmission electron microscopy (TEM) can be very useful for determining the precise lesions in spermatogenic cells. Analyses of cell proliferation and programmed cell death are usually called for in the analysis of a sperm generation defect. When spermatogenesis begins to fail, it usually worsens with time, so analyze the homozygous mutant males at several time points. Finally, if you find a defect in sperm formation, it may be caused by a defect that is intrinsic to the germ cells, the supporting somatic cells of the testis, or even the brain or pituitary. Insights into the primary defect can be obtained by knowing which tissues express the gene of interest. If the gene is expressed in both germ cells and other relevant cell types, a conditional knockout might help to identify the defective tissue (see Section 8.C).

Infertility in females. Infertility in one sex does not necessarily predict infertility in the other. As always, the gene expression pattern of the gene mutated can be a guide. Genes expressed in primordial germ cells, the early indifferent gonad, the gonads, or other reproductive organs of both sexes could potentially affect the fertility of both sexes. Other genes may be exclusive to the gonads or other reproductive organs of one sex or the other.

As a starting point, breed the females to determine if they are fertile. However, if you have generated extra homozygous mutant females and already suspect a potential effect on female reproduction, you can do an initial examination of the ovaries and reproductive organs in parallel with the breeding that can potentially get you to a phenotype more quickly. You should have already determined that the homozygous mutant females look like females and have all of the female-specific organs (see Figures 6.5–6.7). Examine the morphology and histology of the homozygous mutant ovaries, uterus, and oviducts and compare to age-matched controls. In mice, an initial wave of folliculogenesis takes place in a stereotypic pattern soon after birth (Figure 6.12). By 3–5 days after birth, the ovaries contain oocytes that are all arrested in meiosis prophase. By 10–12 days after birth, secondary-stage follicles are present that contain oocytes surrounded by two or more layers of granulosa cells. Between 14 and 24 days, some follicles have progressed to antral stages. Cells closest to the oocytes are called cumulus granulosa cells and cells lining the antrum wall are called mural granulosa cells. By 5–6 weeks, there may be evidence of ovulation (i.e., corpus lutea). Are all of these stages of folliculogenesis present in the homozygous mutant ovary? A block in a specific stage of folliculogenesis will point you in the right direction for finding the primary defect in oogenesis. Check sufficient numbers of homozygous mutant females and sibling controls to determine if the phenotypes that you are observing are only observed in the homozygous mutants.

Mate homozygous mutant females with wild-type males. Set up the matings and wait for a pregnancy or establish timed matings to obtain plugs. The advantage of setting up timed matings is that you can determine if the homozygous mutant females are capable of being plugged. If no plugs are found in the homozygous mutant females, then the chances of obtaining a productive pregnancy are very low. The lack of a plug could suggest a behavioral problem and not necessarily a germ cell defect. Assisted reproduction techniques (e.g., in vitro fertilization) can be used to determine if the ovaries and germ cells are functional (see Section 4.D.3.b). If there are plugs but no pregnancies, then you will need to determine the basis for this infertility.

If you have infertile homozygous mutant females, determine if you have recovered the predicted proportion of homozygous mutant females from the heterozygous intercross matings. Do a statistical analysis to reach a conclusion (see Box 5.1, pp. 72–73). If there are too many females, determine the genetic sex of all of the homozygous mutants (see Box 3.7, p. 40). This will indicate whether you have phenotypic females that are genetically male (XY). The presence of a Y chromosome will cause defects in oogenesis and no functional oocytes will be produced.

If homozygous mutant females do not become pregnant, then use hormones to induce superovulation (MM3) in 3-week-old homozygous mutant females and mate with wild-type male studs. If a plug is present the next morning, collect the oocytes and look under a dissection microscope to determine if they have been fertilized (i.e., they contain two pronuclei; see Figure 5.4B). Culture them overnight to see if they progress to the two-cell stage and, if so, score the percentage of two-cell-stage embryos in comparison with controls. For a more stringent test, continue culturing the embryos to the blastocyst stage (see Box 5.4, p. 82). This can also be done using natural timed matings. If you are able to obtain fertilized oocytes that progress in culture to the blastocyst stage, then the fertility defect is most likely at the stage of implantation or later. If the mutant oocytes cannot be fertilized or they do not progress, this indicates that the oocytes are in some way defective.

6.B.1.c Age-dependent phenotypes

Many genes act to maintain homeostasis after birth. Thus, phenotypes caused by mutations in these types of genes may not become apparent until the mouse ages sufficiently. Similarly, genes that affect the propensity to develop certain diseases may only be evident in older mice.

Tumor formation. The type of gene or its expression pattern may already indicate that the mutation you have generated has a role in tumor formation. To determine if this is true, you will need to age your mutant mice and systematically analyze them for tumor formation. This involves generating sufficient numbers of experimental and control mice, allowing them to age under the same conditions, and determining if, when, and what types of tumors form. Ultimately, you will generate a tumor-free survival curve that plots the number of mice surviving tumor free as a function of time (Figure 6.13). You will also be able to determine the mean latency of tumor detection. In addition, this type of study will provide information on the types of tumors that form, i.e., the tumor spectrum. Design the tumor watch study very carefully because it can take 2–3 years to perform, complete, and analyze. If you do not have a sufficient number of mice with the appropriate genotypes at the beginning of your study, you may not obtain statistically significant (i.e., publishable) data, potentially wasting years of effort.

First, generate a cohort of homozygous and heterozygous mutants as well as wild-type controls that are sex and genetic-background matched. Decide on the number of mice for each genotype that you will need for the study. Publications that include survival curves generally have data for about 25 experimental and 25 control animals. We suggest that you generate 50 homozygous mutants, 50 heterozygotes, and 50 wild-type mice (equal numbers of males and females) because you will lose mice along the way (e.g., a mouse may die on a weekend and its tissues could degrade or be cannibalized before you can get to it). Fortunately, you do not have to generate the mice all at once. They can be accumulated over time. The genetic back-

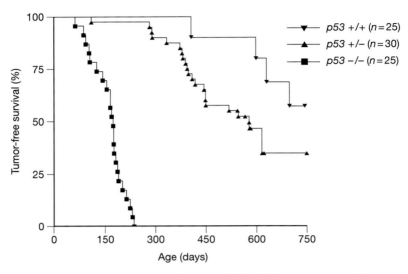

Figure 6.13. Tumor-free survival curves. A cohort of *p53* (*Trp53*) homozygous and heterozygous mutants and wild-type controls was followed over time for survival and tumor formation. (Image provided by Gigi Lozano.)

ground is a very important consideration and should be as consistent as possible to minimize variation (see Section 4.B). The most straightforward choices are an inbred (probably 129) or congenic (perhaps B6) genetic background, or an F_1 between two inbred strains. Most researchers typically cross their chimeras to B6 to generate F_1 hybrids to produce the first generation of heterozygotes. These heterozygotes are then used to produce the F_2 generation to analyze homozygous mutants. Thus, an initial tumor formation study will often be examined on a mixed, F_2 genetic background. An F_2 genetic background can be informative because it contains the entire range of genetic variation between the two inbred strains used. F_3 or backcrosses to one of the inbred strains create genetic backgrounds that may be biased depending on the genetic mixture of the parents used to generate the cohort for the tumor watch.

Once mice are entered into the study at a specific age, monitor them on a regular basis (e.g., twice a week) for obvious tumor formation, moribundity, or death. Sacrifice moribund mice for analysis. Ultimately, this will result in the generation of what is called a tumor-free survival curve (Figure 6.13). For each tumor-bearing animal that is sacrificed, perform a complete histopathological workup (Box 6.4). If the tumor is large, fix a small piece for histology, in situ hybridization, and immunohistochemistry. These tumor samples are precious because they take so long to generate. Thus, make sure that the fixative you use is compatible with the different types of analyses you might perform. If the tumor is large enough, snap freeze it (dry ice or liquid nitrogen) for protein and nucleic acid analyses (e.g., loss of heterozygosity studies). If you are not a histopathologist, collaboration with one could be valuable. The tumor watch should be carried out until the mice are at least 2 years of age before concluding that there is no effect of a homozygous or heterozygous mutation. However, a strong effect on tumor formation could be detected in much less time.

> **Box 6.4** **Necropsy Guide for the Collection of Tissues from Mice With or Without Tumors**
>
> *Time.* Perform necropsy as soon as the mouse is euthanized or is found dead. If the necropsy cannot be performed immediately, refrigerate the body. Never freeze the body because this will cause tissue distortion. However, specific tissues may be frozen if frozen sections are needed.
>
> *Fixation.* Fix tissues minimally at a 1:10 ratio of tissue to fixative. Once fixed, the tissues do not need to be refrigerated.
>
> *Size.* Ideally, tissue samples should be no thicker than 5 mm for good penetration of the fixative. The length and width can be greater. Sometimes you may have no choice but to fix a large sample (e.g., the entire head).
>
> *Tissue samples.* Generate a list of tissues that will be collected from every mouse. You can never take too many samples because you cannot predict whether you will have to go back to a certain tissue for more analysis.
>
> *Tumor description and documentation.* Describe the location, size, shape, pattern, amount, consistency, involvement, and color of each tumor. Take pictures of the tumor in situ and dissected away from the body. Take tissue samples from the tumor and nearby control tissues.

Other age-related pathologies. Many other types of pathologies can develop as mice age, including skeletal, cardiovascular, and neurological. The approach to studying these mutant phenotypes is generally the same. Sufficient numbers of age- and sex-matched mutants and controls are set aside to age for various amounts of time and examined for the development of the pathology of interest. Here we discuss the analysis of age-related skeletal pathologies as a general example. Because of the long time frame, aging studies are generally only done when you have sufficient reason to suspect that an age-related pathology will develop based on the properties of the mutated gene.

Age-related skeletal pathologies include osteoporosis (bone loss), osteopetrosis (excess bone formation or lack of remodeling), and arthritis that can become progressively more severe as homozygous mutant mice age. Bone formation is a balance between osteoblast and osteoclast function. Accordingly, these are the two primary cell types to consider when presented with bone formation pathologies. Arthritis is a disease of the joints. Therefore, the joints would be the focus of an investigation of arthritis. You may notice these pathologies initially because they can lead to overt abnormalities such as kyphosis (curvature of the spine), reduced movements, or swelling of the joints.

The analysis begins with a documentation of the intact live animals or the structures of interest (e.g., the limbs). Photograph the homozygous mutants and age- and sex-matched controls while they are either awake or anesthetized. Make sure to include a size marker in the image (e.g., a ruler). To assess skeleton structure without sacrificing animals, take X-ray pictures of anesthetized mice (Figure 6.14). These X-ray images can provide sufficient resolution to determine general bone density and other bone abnormalities (e.g., fractures). A next step might be to generate a skeleton preparation (MM3) to grossly examine the bones and joints in more detail. Bone histopathology will provide cellular and tissue organization details.

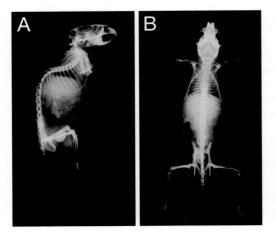

Figure 6.14. Contact X-ray images of skeletons. (*A*) Lateral view of wild-type 3-month-old female mouse. (*B*) Dorsal view of same mouse. Repositioning the mouse is used to show specific regions of the skeleton. (Image courtesy of Yuji Mishina and Richard Behringer.)

Once you have an initial indication that you are dealing with an age-dependent skeletal pathology, you will next need to perform a comprehensive temporal study of the development of the pathology. Set aside cohorts of homozygous and heterozygous mutants and wild-type controls that will be analyzed at various ages. Remember that there may be differences in the phenotypes of males and females. Thus, it is necessary that you also set aside sex-matched controls. Because these types of studies can potentially take a long time to perform, make sure that you have also set aside sufficient numbers of animals to age to generate statistically significant data.

6.B.2 No visible mutant phenotypes

(Notice that we do not use the term "no phenotype" because it is highly unlikely that the mutation of a functional gene will be completely without effect. More likely, it is a matter of knowing how to look for the phenotype.) To arrive at this section, you found that homozygous mutants, both male and female, are born at Mendelian frequency, survive to adulthood, are fully fertile, and have no obvious abnormalities. Unless you have very strong reasons to believe that the mutation will only result in a late-onset phenotype, you probably will not want to wait until the mice are 2 years old before exploring other avenues. You could set up a tumor watch (see previous section), if tumors are a possibility, or set aside a cohort of mice to age, but at the same time, try some experiments that might reveal a mutant phenotype. In this section, we delve more deeply into less visible phenotypes, such as behavioral modifications, and also present the mutants with various challenges that might reveal a phenotype. In following up these lines of investigation, you can be selective depending on the expression pattern of the gene or what you know about the protein and its potential activity. Before proceeding, however, it is worth considering some trivial explanations for your mutant not having a visible phenotype. Perform a few control experiments to verify that your homozygous mutants really are null for the gene under study and that you are recovering the expected numbers of homozygous mutants.

The following are some possible causes for no visible phenotype:

1. The mice are not homozygous mutants. Somehow, an error occurred in genotyping. Check your Southern blots or PCR results. If you are using Southern analysis, make sure that you are using the correct probe (i.e., sequence it). Recut the tails of the mice and re-genotype.

2. If you designed a loss-of-function allele, you may not have completely inactivated the gene. Review your knockout strategy (see Section 2.B). Because this is the first time that you will have a homozygous mutant mouse, you can perform a rigorous experimental test to determine if you have truly deleted the desired sequences. Perform a Southern analysis of homozygous mutant DNA (tail or other tissue), using a probe within the region that has been deleted. The predicted wild-type band should be absent. Strip the membrane and rehybridize it with an irrelevant probe as a positive control for the presence of DNA. If you have not deleted all protein-coding exons, perform reverse transcriptase (RT)-PCR or northern analysis using probes that are 5′ and 3′ of the targeted deletion to determine if any transcripts are produced from the mutated locus. If you have antibodies to the gene product, perform a western blot or immunohistochemistry on homozygous mutant tissues to determine if any protein or partial protein is generated.

3. The nontrivial possibility is that homozygosity for the null mutation results in no overt abnormality. Go to Sections 6.B.2.a, 6.B.2.b, 6.B.2.c, or 6.B.2.d as seems appropriate for the expression pattern of your gene.

HELPFUL HINT

It is possible that you will recover homozygous mutants that are viable and fertile with no obvious defects but fewer of them than the predicted Mendelian frequency. You may have a variably penetrant mutant phenotype such that some homozygous mutants are viable with no obvious abnormalities but some die before birth and thus are not detected. Examine the ratios of homozygous mutants obtained relative to heterozygotes and wild type, using a larger sample than before (see Box 5.1, pp. 72–73). If you find a statistically significant underrepresentation of homozygous mutants after birth, establish timed matings between heterozygotes to determine when the homozygous mutants are being lost and analyze the prenatal phenotype (see Chapter 5).

6.B.2.a Tests of the senses

Here we present some simple tests to determine if your homozygous mutant mouse can perceive fundamental senses. If these simple tests reveal a defect, investigate more sophisticated measures of the specific senses or seek out an expert collaborator.

Vision. How can you determine if a mouse has normal vision? Unfortunately, there is no simple, definitive way to test for vision in mice. A first attempt at assessing normalcy of the eye would simply be a gross dissection and histological analysis in comparison with wild-type controls. This will at least indicate any major abnormalities in the eye (e.g., abnormal retina). The simplest means is to assess response to light by observing if the pupils constrict when exposed

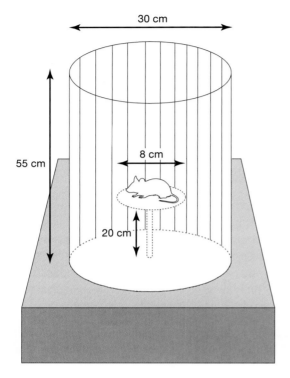

Figure 6.15. Diagram of visual tracking drum apparatus. (Redrawn, with permission, from Thaung et al. *Neurosci. Lett. 325:* 21–24 [2002 © Elsevier].)

to light (e.g., from a small flashlight) and dilate when the light is removed, although this does not test visual function. Below are two other functional tests to assess vision in mice, both requiring specialized equipment and varying amounts of technical expertise.

One method uses an apparatus called a visual tracking drum (Figure 6.15). The visual tracking drum is a cylinder approximately 1 foot in diameter and 20 inches tall that can rotate at two revolutions per minute. Cards with black and white vertical stripes are placed on the inside of the drum. The mouse is placed on a central, stationary raised platform. The top of the drum is open to monitor the head movements of the mouse. If the mouse can see the pattern on the inside of the drum as it rotates, it will follow the movement of the pattern with its head. If its vision is severely compromised, it will not exhibit this "head-tracking" behavior. This method does not require training the mouse because head-tracking behavior is a reflex response.

If you need more detailed measurements of vision, consider collaborating with a vision expert to generate an electroretinogram (ERG) for your mouse. ERG recordings measure the eye's electrical response to a flash of light. Responses are recorded using an electrode placed on the surface of the eye while the mouse is anesthetized. The ERG is usually generated by an instrument called a ganzfeld stimulator, which assesses the state of the entire retina.

Hearing. A simple test of severe hearing loss (i.e., deafness) is to use a click box to generate a sound of specific intensity to test for an acoustic startle response or the Preyer reflex, a physical response to an auditory stimulus. The click box generates a brief 20-kHz tone at a 90-db sound pressure level (SPL) when held 30 cm above the mouse. The Preyer reflex can be strong (the mouse jumps) or subtle (a backwards flick of the ears). This test can identify both PNS and

> **ALERT**
>
> *Many of the common inbred mouse strains carry a mutation called* retinal degeneration *(Pde6b) that causes postnatal loss of photoreceptors in the retina, resulting in vision loss. Substrains of 129 and C57BL/6 mice are wild type for* Pde6b. *Some common strains that are* Pde6b/Pde6b *include C3H and CBA. Thus, if you are studying genes that may influence the eye, be sure to check the* Pde6b *genotype of your genetic background or you may end up chasing a phenotype that is not specific to your engineered mutation.*

CNS defects. Because mice respond to the ultrasonic vocalizations of their offspring for such behaviors as recovering scattered pups, poor mothering skills can indicate deafness.

Smell and taste. Tests for smell and taste are quite specialized. It is likely that you will not be concerned about these types of phenotypes unless your gene is expressed in the nose, tongue, or relevant regions of the brain and/or this is your primary area of expertise. However, a simple test to detect the absence of a sense of smell is to bury a piece of smelly cheese or chocolate in the home cage and measure the time it takes homozygous mutants to discover it compared with control mice. As with hearing, mice depend to some extent on olfactory cues for the recognition of their offspring. The loss of the sense of smell could lead to bad parenting, as evidenced by pups scattered out of the nest or even cannibalized.

Touch. The perception of touch allows a mouse to interact with its environment and avoid injury. In the mouse, touch perception is routinely assessed using the Von Frey filament test. The mouse is placed on a wire mesh platform and Von Frey filaments (fibers of different diameters) are inserted through the mesh from below, touching the footpads. The force of the filament on the skin increases as you continue to push the probe until the filament bends. Once the filament bends, the force will be maintained, not increased. Mice will withdraw their paws when they perceive the touch. A threshold for withdrawal is determined by using filaments of increasing diameter.

Nociception is the perception of pain. There are several different types of pain. Acute pain can be a "good" pain because it protects an organism from injury by outside noxious stimuli, such as fire, hot water, etc. In rodents, two simple tests for heat-induced pain are the tail flick test and the hot plate test. The tail-flick test measures reflex spinal nociception. The mouse is restrained in a tube and the tail is either exposed to a radiant heat source or it is immersed in water at 52°C. The amount of time it takes for the mouse to flick its tail away from the heat source or remove it from the water is measured. The cutoff time is 10 seconds. The hotplate test measures supraspinal nociception. By placing the mouse on a hot plate set at 52°C, a wild-type mouse will either lick its hind paws or jump off of the hot plate within seconds, whereas a mutant mouse may have a delayed response or no response at all. The maximum exposure of the mouse to the heat is 30 seconds.

Chronic pain perception is measured using an inflammation-induced paradigm, the injection of a dilute solution of formalin or Freund's adjuvant into the mouse's paw. Inject 15 μl of a 5% formalin solution subcutaneously into the ventral surface of one hind foot. A wild-type mouse will feel pain induced by the formalin. The behavioral measurements of this type of pain include favoring, lifting, shaking, and licking the paw. These measurements are taken periodically over approximately 30 minutes. Mice with defects in the perception of nociception will not exhibit these behaviors or will show these behaviors less frequently than controls.

Before performing any of the above tests, be sure to obtain the necessary approvals from your institutional animal care and use committee, which can provide regulations for any tests of pain perception.

6.B.2.b Neurological tests of balance and coordination

Careful observation of mice in their home cage can sometimes reveal behavioral or neurological problems. For example, circling or twirling behavior or a head tilt can indicate vestibular problems. Even though a mouse appears normal by visual inspection, it may still have a neurological defect that can be detected using some very simple tests. Below are a few that can be easily performed without complex and expensive equipment.

Tail suspension test. The tail suspension test is a simple test that can reveal if a mouse has some kinds of neurological defects. It is indicative of a developmental defect, as the response changes with age, but it cannot distinguish between a CNS, PNS, or combined abnormality. Hold the mouse by the tail suspended over the cage. A wild-type mouse will spread its limbs outward, whereas mice with certain neurological defects will hold their limbs into their body, so-called "huggers" (Figure 6.16). Wild-type mice at 12–14 days postpartum also show this hugging behavior but they eventually outgrow it.

Balance and motor coordination. A simple test of balance and motor coordination is the rotarod test. The rotarod is a machine with a dowel held in a horizontal position that rotates with increasing speed over time (Figure 6.17). A wild-type mouse can be trained to stay on the rotarod for a certain period of time. However, it will eventually fall off as the speed is increased.

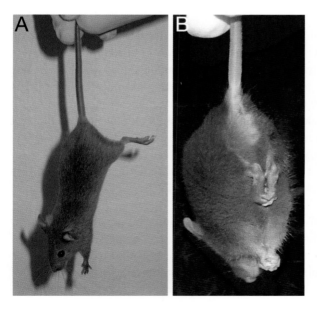

Figure 6.16. Tail suspension test. Wild-type mice will spread their limbs when suspended by their tails (*A*). However, mice with neurological defects may grasp their limbs to their bodies, so-called "huggers" (*B*). (*B* courtesy of Andy Salinger and Monica Justice.)

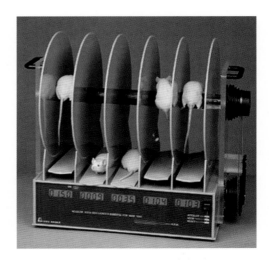

Figure 6.17. Rotarod apparatus. (Image courtesy of myNeuroLab.com, St. Louis, MO.)

This test is quantitative because you score the number of seconds the mouse can stay on the rotarod. Mice with defects in balance and coordination, possibly indicative of a cerebellar defect, will fall off the rotarod quicker compared to wild-type mice.

Strength. We have all had the experience of a mouse clinging to the wire lid of a cage as the lid is lifted, sometimes leading to concern about missing mice. This trait of mice can be used as a simple measure of grip and strength. Place a mouse on top of a wire cage lid, shake it several times to make sure the mouse is gripping, then invert the lid over the cage at about 25 cm. Normal mice can hold this upside-down position for several minutes. Measuring the time it takes a mutant mouse to drop compared to wild-type mice could reveal a neuromuscular defect. Cutoff time is 60 seconds.

Gait analysis. The gait of a mouse can be informative. Wild-type mice have a regular, alternating gait. Variation from this pattern can indicate ataxia or a morpholigcal defect that causes an abnormal gait. First, watch how the homozygous mutant walks around the cage and compare this to wild-type mice. Are there any obvious differences? To document the gait of a mouse, dip the feet of your mouse in nontoxic ink and have the mouse walk on a large piece of filter paper before the ink dries. You could use different colors of ink for the forelimbs and hindlimbs. We find it useful to create a chute through which the mouse can walk, which gives a nice visual record of the gait in one direction (Figure 6.18). If you identify an abnormal gait in the homozygous mutants, a video record is useful for further analysis.

Swim test. The ability to swim is a common test of vestibular function of the inner ear. Mice with vestibular defects cannot maintain their orientation to gravity when not in contact with a solid surface. Fill a large container (e.g., a mouse cage or aquarium) with tepid water. Hold the mouse by the tail and lower it into the water. A wild-type mouse will swim ("dog paddle"). Mice with vestibular defects will be unable to swim or to remain at the surface of the water. The inability to swim could also be a result of defects in muscles, skeleton, limbs, and nerves. Be ready to rescue mice that cannot swim and dry all the mice off in a warm cage.

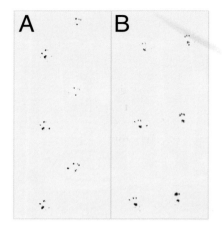

Figure 6.18. Footprints of mice to show gait. Ink is applied to the hindlimbs of wild-type and mutant mice. The mice then walk on white filter paper to leave their tracks. Wild-type mice (*A*) have an alternating gait. In this example, the mutant (*B*) hops.

6.B.2.c Genetic challenges of viable, overtly normal homozygous mutant mice

A mouse homozygous for a null mutation is, by definition, deficient for a specific gene product. Accordingly, even if it is viable and appears normal, one can consider the genotype of the mouse to be sensitized. Thus, when challenged, the homozygous mutant may yield an abnormal phenotype. The complexity of the mammalian genome is likely the explanation for a high frequency of overtly normal homozygous mutant mice. Therefore, think about the types of challenge experiments to pursue, even at the beginning of each project, based on what you know about the gene and its protein.

Conceptually, two types of challenges can be employed: genetic and environmental. Both types of challenges can yield information about the role of the gene that is mutated in development and homeostasis. Of course, the more you know about the gene in question, the easier it will be to decide what type of challenge to perform. A genetic challenge to a mutation is basically a genetic interaction experiment (Section 4.E.3). The gene that is mutated may have redundant functions with another gene. Alternatively, it may be compensated by the up-regulation of another gene. Adding an additional mutation within the same biochemical pathway may compromise that pathway, leading to a mutant phenotype. A full discussion of crosses to generate double mutants is provided in Section 4.E.3. The expression of transgenes on a homozygous mutant genetic background may also yield a phenotype specific to the targeted mutation.

Genetic background. As we have mentioned before (Section 4.B), the genetic background on which a mutation is studied can have a great influence on the penetrance and expressivity of the mutant phenotype. You can move the mutation onto any inbred genetic background by generating a congenic line (see Box 4.4 on p. 54 and Figure 4.2). The choice of the genetic background can be based on the well-known characteristics of the specific inbred background. You can also move the mutation onto an outbred genetic background to expose the mutation to genetic diversity. An extension of this idea is to cross mice with the mutation to a different species or subspecies of mouse (e.g., *Mus spretus* or *Mus castaneus*). The logic behind this strategy is that a different species or subspecies of mouse will have many more genetic differences relative to a different inbred or even outbred strain to modify the expression of a mutation. If you try this, keep in mind that *Mus musculus/Mus spretus* F_1 interspecific hybrid males are sterile. Fortunately, the F_1 interspecific hybrid females are fertile. There are no sterility concerns with *Mus musculus/Mus castaneus* F_1 intraspecific hybrid males and females.

Crosses with a transgenic strain. Typically, transgenic mice generated by zygote injections express gene products at high levels in ectopic tissues. These gain-of-function experiments can provide a genetic challenge to a homozygous mutant mouse that is otherwise phenotypically normal, causing it to yield a mutant phenotype. For example, consider a gene that is suspected of being involved in cancer that, when mutated, results in homozygous mutant mice that are viable and overtly normal. One way to challenge these mice genetically would be to use a transgenic mouse line that expresses an oncogene that induces tumor formation. In mice homozygous for the engineered mutation who also carry the transgene, tumor formation may be enhanced or suppressed, revealing a role for the engineered mutation in regulating tumorigenesis.

6.B.2.d Environmental challenges of viable, overtly normal homozygous mutant mice

A challenge to a viable, overtly normal homozygous mutant does not necessarily have to be a genetic one. It is also possible, depending on the expression pattern and the putative role of the gene in question, to use environmental challenges to elicit a mutant phenotype from an otherwise normal homozygous mutant. The many types of environmental challenges that can be applied to a mutant include a variety of physiological stresses. Alterations in metabolism could be challenged by feeding the mice a special diet (e.g., high fat). A high-fat diet could also be used to stress the cardiovascular system by increasing the formation of fat deposits on arterial walls. Chemicals (e.g., carcinogens) and radiation are common challenges to induce tumor formation or reveal defects in DNA repair and genome stability, respectively. A challenge of the immune system can be performed by exposing the mutants to various pathogens (e.g., bacterial). Remember that many of these types of environmental challenges can also be applied to cells derived from the homozygous mutant (e.g., mouse embryonic fibroblasts) to reveal cellular phenotypes.

CHAPTER 7

Dominant Effects

STRICTLY SPEAKING, A DOMINANT MUTATION is one that produces a specific phenotype in heterozygous mice and the same phenotype in homozygous mutants. A semidominant mutation is one in which there is a specific phenotype in heterozygotes, but an even more severe version of this phenotype in homozygous mutants. Here, for the practical purpose of mutation analysis, we consider a dominant effect to be any phenotype seen in a heterozygous mouse. When more is known about how and why a phenotype appears and whether the homozygous effect is more severe, the mutation can be classified as dominant or semidominant. Dominant effects of a mutation can appear at almost any stage in the process of making a targeted mutation or at any stage of development either in targeted, induced, or spontaneously occurring mutations. This chapter provides additional information about dominant effects and how to deal with them. If you have a mutant mouse with a dominant phenotype, you might want to take a detour now through Chapter 4 for hints on breeding in general and Section 4.D.3 for specific tips on breeding dominant fertility mutants.

7.A How Dominant Effects Come About

The dominant effects of a mutation can occur in several ways depending on the nature of the genetic lesion (Figure 7.1). If a mutation results in a null allele such that there is no gene product produced, a heterozygous phenotype could be the result of haploinsufficiency for the gene product. This phenotype implies that there is a dose-sensitive requirement for the gene product and that the haploid amount is not sufficient to produce a wild-type phenotype. Conversely, a mutation may result in a hyperactive protein (a hypermorph), which could also be detrimental. Other

Chapter opening figure courtesy of Debbie Chapman.

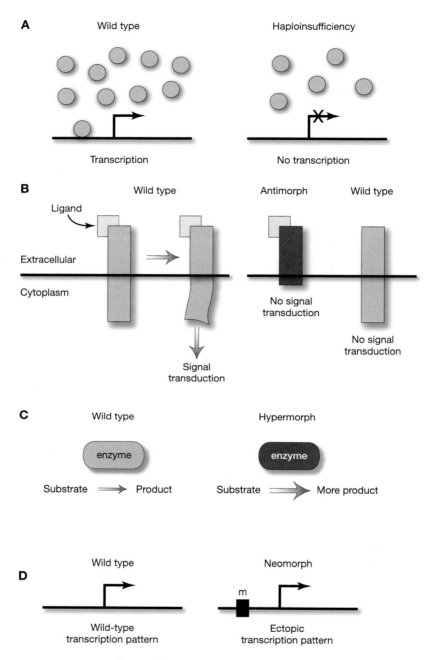

Figure 7.1. (*See facing page for legend.*)

mechanisms for obtaining a dominant effect would be a mutation that results in the production of an altered protein that in some way interferes with the function of the wild-type protein (an antimorph or dominant negative) or an altered protein or expression pattern that produces a new activity, possibly unrelated to the function of the wild-type protein (a neomorph). In these cases, it is not the lowered level of wild-type protein that causes the phenotype, but rather the presence of an abnormal protein or expression pattern. For null alleles, there are also several special cases where the loss of a single allele can result in the complete lack of a gene product rather than just haploinsufficiency, and in these situations, a mutation is quite likely to act as a dominant.

1. *A mutation in an X-chromosome-linked gene.* In males (or male embryonic stem [ES] cells) with only one X chromosome, a mutation in an X-linked gene results in the only copy of the gene being mutated. In females, most tissues undergo random X-chromosome inactivation and thus, in females with an X-linked mutation, half of the cells will have only one wild-type allele functional and the other half will have no wild-type alleles functional, creating a mosaic of cells with and without the gene product (Figure 7.2A,B). In certain embryonic tissues, such as the trophectoderm, primitive endoderm, and their derivatives, there is preferential inactivation of the paternal X chromosome. In these tissues, an X-linked mutation will result in an effective null in heterozygous females, i.e., a dominant effect if the mutant allele is inherited from the mother but no effect if it is inherited from the father (Figure 7.2C,D; see also Figure 1.1).

2. *A mutation in a Y-chromosome-linked gene.* Just like mutations in X-chromosome-linked genes, Y-chromosome-linked gene mutations will show dominant effects in males (or male ES cells) because the only copy of the gene is mutated. A likely consequence of a Y-chromosome gene mutation is male infertility because many of the genes on the Y chromosome are involved in spermatogenesis.

3. *Imprinted autosomal genes resulting in an effective null with only a single copy of the mutated allele.* These are genes that are epigenetically inactivated depending on whether they are inherited through the male or female germ line, i.e., from the father or the mother. For example, if a gene is inactivated by imprinting when it is derived from the father, a null mutation coming from the mother will result in an embryo with a null phenotype, even though it is genetically heterozygous (see Figure 1.2).

Obviously if you know that your gene is X linked or imprinted, these possibilities should have already been considered (Section 1.A and Figures 1.1 and 1.2). If the gene is not X linked and does not reside in a chromosomal region already known to be subject to imprinting, it could still be imprinted. Imprinting is easily checked by determining whether the mutation has the same dominant effect when inherited from the mother as it does when inherited from the father. For special considerations for dealing with imprinted genes, go to Section 7.E. For dominant effects appearing in ES cells, go to Section 7.B. For dominant effects in chimeras, go to Section 7.C. For dominant effects in heterozygous offspring of chimeras or in mice heterozygous for spontaneous or induced mutations, go to Section 7.D.

Figure 7.1. Mechanisms of action of dominant mutations. (*A*) Haploinsufficiency. In this example, wild-type levels of a transcription factor are required to achieve a critical concentration to activate transcription of a target gene. When there is only one dose of the gene, the level of protein is too low (i.e., the concentration is too low) to activate target gene transcription. (*B*) Dominant negative or antimorph. In this example, binding of a ligand to a wild-type transmembrane receptor induces a conformational change of its cytoplasmic domain that causes a signal to be transduced. A mutant version of the receptor (darker shading) can bind the ligand, and if it is at limiting concentration, sequester it from the wild-type receptors. Because the mutation truncates the protein, it lacks the cytoplasmic domain and cannot transduce a signal. (*C*) Hypermorph. In this example, a wild-type enzyme acts on a substrate to catalyze the production of a product. The mutant protein can catalyze this reaction more efficiently, leading to a greater production of the product. (*D*) Neomorph. In this example, a wild-type allele is expressed in a specific pattern. The mutation (*black box* [m]) causes a change in the tissue specificity of transcription, leading to a new, ectopic site of expression.

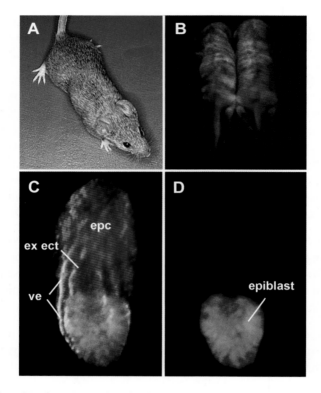

Figure 7.2. Examples of X-chromosome inactivation. (*A*) *Tabby* is an X-chromosome-linked mutation that affects hair, tooth, and endocrine gland development. In heterozygous females, random X-chromosome inactivation on an agouti genetic background leads to coat color variegation. Note the fine transverse stripes of light-colored normal and dark tabby hair. (*B*) Newborn female mice carrying one copy of an X-linked green fluorescent protein (GFP) transgene. Random inactivation of one X chromosome results in a mosaic fluorescent/nonfluorescent pattern in the skin. (*C,D*) Nonrandom inactivation of the paternally derived X chromosome in extraembryonic tissues. This 7.5-day postcoitus (dpc) female embryo inherited an X-linked GFP transgene from its father. There is a mosaic of GFP-expressing and nonexpressing cells in the embryonic portion of the embryo, the epiblast (as seen by fluorescence microscopy in both *C* and *D*), indicating random X inactivation, whereas the extraembryonic region is nonfluorescent (seen as orange in *C*, which is an overlay of bright-field and fluorescence images) due to preferential inactivation of the paternally derived X chromosome in the extraembryonic tissues. (epc) Ectoplacental cone; (ex ect) extraembryonic ectoderm; (ve) visceral endoderm. (*A* Courtesy of the Mouse Genome Database, http://www.informatics.jax.org [The Jackson Laboratory, Bar Harbor, Maine]; *B–D* courtesy of Kat Hadjantonakis.)

7.B Dominant Effects in ES Cells

We have pointed out the unlikelihood of a mutant being unrecoverable in ES cells due to a dominant mutant effect (Section 3.A.e), although it is still a formal possibility. Because the outcome of this problem would be the lack of targeted ES cell colonies, the same as if there were no targeting events at all, this will be virtually impossible to prove because there will never be any cells to study. If the gene is X-chromosome linked and you have been using an XY ES cell line without success, you might be able to circumvent the problem by using an XX ES cell line for gene targeting. If the gene is Y-linked, this will not be an option. Targeting a gene in an imprinted region should not create this type of problem because only one of the two alleles is likely to be imprinted, and furthermore, imprints appear to be lost altogether in some ES cell lines.

If the gene is neither sex-chromosome-linked nor imprinted, a genuine dominant effect that prevents the recovery of targeted ES cell clones would have to affect a basic cellular function, such as proliferation, metabolism, adhesion, etc., essential for ES cell survival in tissue culture. In the case of a null mutation, it is unlikely that a truly basic cellular function would be subject to such acute dose sensitivity, but if the mutation is a dominant negative (antimorph) or neomorph, the aberrant protein might be able to wreak havoc on critical basic cellular functions. In either case, a possibility for further study would be to make a different type of mutation, e.g., a conditional null or a hypomorph (see Section 8.C) or to use a so-called rescuing transgene (Box 7.1 and Section 8.B).

7.C Dominant Effects in Chimeras

Another unlikely but formally possible situation is a dominant effect that does not affect ES cell growth, but causes the death of chimeric embryos (Section 3.B.1.c). This could indicate a non-cell-autonomous effect such that the presence of the mutant cells is actually detrimental to the development of the embryo as a whole. It could also be the result of the ES cells initially making very large contributions to critical tissues in the chimera, but failing later due to the effects of the mutation, causing death of the chimera. However, a much more likely outcome of haploinsufficiency would be poor contribution of the heterozygous ES cells to chimeras or contributions only to specific (unaffected) cell and tissue types (Section 3.B.2.e). On the other hand, for example, a neomorph or dominant-negative (antimorph) mutation of a critical signaling molecule might well disrupt the development of a chimera. In either case, further analysis could be accomplished by making a constant supply of chimeras and following the analytical steps outlined in Chapter 5 for homozygous mutants to pin down the time and cause of death of the chimeras. For this analysis, the inclusion of a cell marker specific to either the ES cells or the embryonic cells is essential for recognizing chimeras and determining the tissue distribution of the heterozygous ES cells throughout development (see Box 8.1, p. 180).

Any number of genes might have dominant effects that could be seen in neonatal or adult chimeras. We mentioned a situation that could even mask chimerism (Section 3.B.2.d). If the mutation you made affects coat color, it is possible that a dominant effect would show up in chimeras if the effect could be seen on the specific genetic background of the ES cells. For example, in black ES cell↔albino embryo chimeras, a dominant mutation causing the death of melanoblasts could result in a cryptic chimera with an all-white coat. Similarly, in agouti↔nonagouti chimeras, a dominant effect on the agouti signaling pathway might affect the agouti pattern of chimeras. The severity of a dominant phenotype in chimeras will usually be correlated with the level of contribution of ES cells to the affected tissue. If the effect is not too severe and does not affect the reproductive capacity of the chimeras, the mutation should still be recoverable from the germ line of the chimeras and can then be studied in heterozygous or homozygous mutant mice, as outlined in Chapters 5 and 6.

If, however, the dominant effect prevents germ-line transmission of the ES-derived genotype, it will be harder to recover the mutation and establish it in the germ line initially; it will be the same as if there were no germ-line transmission. Any dominant effects limited to

168 | Chapter 7

> **Box 7.1** Transgene Rescue of Mutant ES Cells, Chimeras, or Embryos
>
> A dominant mutation that causes an ES cell-lethal, male-sterile, or embryo-lethal phenotype can be hard to study because you will not recover targeted ES cell clones, generate germ-line-transmitting male chimeras, or obtain viable heterozygous mutants after birth, respectively. To study these types of dominant mutations, one option is to use a transgene, e.g., a wild-type copy of the gene being mutated, to rescue the mutant phenotype. However, hypermorphic and neomorphic mutations cannot be rescued with a wild-type transgene, and transgene rescue depends on wild-type ES cells being tolerant of extra copies of the gene in question. If you suspect a dominant, ES cell-lethal phenotype, you can introduce a wild-type copy of the gene of interest into wild-type ES cells and then use gene targeting to mutate an endogenous allele to recover heterozygous mutant cell lines that are rescued by the wild-type transgene (Figure 7.3A). If heterozygous cell lines are rescued, this provides formal proof of a dominant, ES cell-lethal phenotype and might encourage you to make a hypomorphic allele for further study.
>
> If you suspect a dominant male-sterile or embryo-lethal phenotype, you can introduce a wild-type copy of the gene of interest into already targeted ES cells. Generate male chimeras using the transgene-containing targeted ES cells and breed them with wild-type females. The transgene may rescue a dominant male sterility in the chimeras, leading to germ-line transmission and the recovery of heterozygous progeny with or without the transgene (Figure 7.3B). The male progeny without the transgene can be used to study the dominant effects of the mutation on spermatogenesis, whereas the mutation itself can be maintained by breeding heterozygous females or the heterozygous male progeny carrying the transgene. A transgene might also rescue a dominant embryo lethality. In this case, only progeny carrying the transgene will be viable (Figure 7.3C). Use the heterozygous progeny with the rescuing transgene to breed with wild-type mice. Provided that the integration of the rescuing transgene in not linked tightly to the endogenous locus, segregation of the genes will provide nontransgenic heterozygous mice in the resulting litters that can be used for analysis of the dominant embryo-lethal mutant phenotype. Of course, you will have to genotype for both endogenous alleles and the rescuing transgene to find the desired embryos.
>
> The rescuing transgene can be designed so that it is under the control of its endogenous regulatory sequences or heterologous sequences, depending on the situation. Although conceptually simple, in practical terms it is not always clear which endogenous or heterologous sequences will be most effective for transgene expression in the correct temporal and spatial pattern. Rarely are the endogenous sequences that regulate gene expression thoroughly characterized. In addition, a heterologous regulatory element may not be sufficient for relevant transgene expression, or it may lead to overexpression and a gain-of-function phenotype.
>
> **Figure 7.3.** Transgene rescue of dominant phenotypes. (*A*) Dominant ES cell lethality. A wild-type copy of the gene is electroporated into ES cells before gene targeting. (*B*) Dominant effect on spermatogenesis in chimeras. A rescuing transgene introduced into heterozygous ES cells results in germ-line transmission by allowing spermatogenesis in the chimeras to proceed. (*C*) Dominant embryonic lethality. The rescuing transgene, introduced into heterozygous ES cells, allows heterozygous embryos that carry the transgene to survive. (+/+) Wild type; (+/m) heterozygous mutation; (tg) transgene. (*See facing page.*)

gametogenesis in one sex or the other could be circumvented by breeding the mutation through the germ line of the other sex and then producing sterile offspring of the affected sex for a gametogenesis study. However, in the case of a dominant affect on spermatogenesis (Section 3.C.2.d), or a haploid effect on sperm function (Section 3.C.3.b), the initial problem of lack of germ-line transmission (assuming that you are using an XY ES cell) will require getting the mutation into the female germ line either by breeding from female chimeras or by repeating

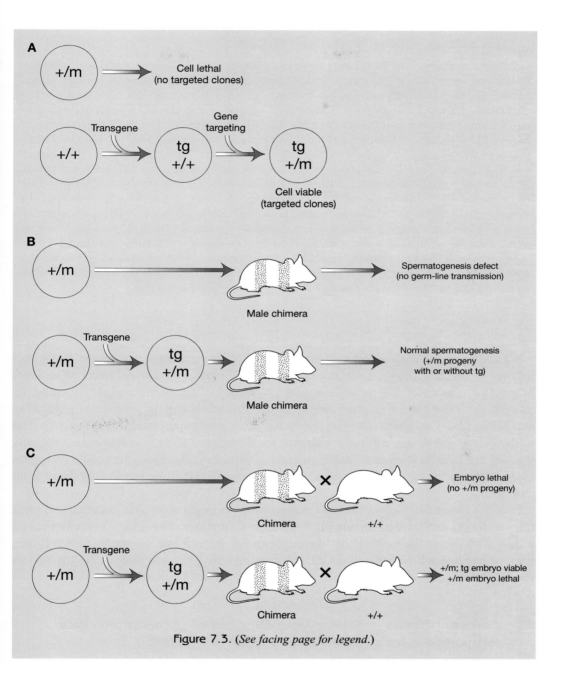

Figure 7.3. (*See facing page for legend.*)

the gene-targeting experiment using a female cell line, and hoping that the effect on gametogenesis does not extend to oogenesis. Alternatively, you could try a rescuing transgene (Box 7.1) or isolating an XO ES cell line from the targeted XY ES cells (see Box 7.2). Because most gene-targeting experiments make use of XY ES cells, any dominant effects on oogenesis will usually not be revealed until the mutation is safely in the germ line (see Section 7.D.2), in which case it can be maintained by breeding through the male.

> **Box 7.2** Isolating XO ES Cells from XY ES Cells
>
> Most of the commonly used ES cell lines are genetically male (XY) and, consequently, most germ-line transmission of the ES cell genotype is through male chimeras. In some situations, you may require transmission through female chimeras but only have XY ES cells available. XO mice develop as fertile females. Thus, an alternative to working with XX ES cells is to isolate XO ES cells from an existing XY wild-type or mutant ES cell line. XY ES cells spontaneously lose the Y chromosome to become XO at about a 1% frequency in vitro. To isolate an XO ES cell line, plate XY ES cells at clonal density using standard ES cell culture methods (MM3).* After 7–10 days, pick 300–400 colonies into 96-well plates for expansion and then split to freeze down master plates and generate duplicate plates for genotyping. The presence or absence of the Y chromosome can be assessed by polymerase chain reaction (PCR) genotyping with Y-chromosome-specific genes such as *Sry* or *Zfy* or by Southern analysis using a mouse Y-chromosome-specific probe such as Y353/B (see Box 3.7, p. 40). You may want to check the Y-chromosome status of your fibroblast feeders, because if they are XY, they may contaminate the ES cells, especially if you are using PCR for genotyping. If this is a problem, ES cells can be purified by brief serial passages on tissue culture dishes because feeder cells adhere much more quickly than ES cells. Once you have isolated multiple XO clones, they can be used to make chimeras, at least 50% of which will be females with a chance of germ-line transmission.

If a dominant gametogenesis effect produces sterility in both sexes, one means of getting a mutation in ES cells into the germ line of mice is to produce a conditional mutation in which the gene remains functional throughout gametogenesis but can be mutated at will using, for example, an inducible Cre to study the gametogenesis phenotype in subsequent generations (see Section 8.C). Alternatively, you might try using a rescuing transgene (Box 7.1). This involves electroporating a wild-type transgene for the targeted allele into the targeted ES cells, selecting for random integrants, and then making chimeras. The transgene should rescue the mutant effect, allowing the targeted endogenous mutant allele to be transmitted through the germ line.

Finally, a dominant mutant effect on sex determination might show up in chimeras and result in feminization, even if chimeras are made with XY ES cells (Section 3.C.1.c). Alternatives for dealing with this problem would be to retarget an XX ES cell line, isolate XO ES cell lines from the targeted XY ES cells to get the mutation into the germ line through the female (see Box 7.2), or make a conditional allele (Sections 2.B.4 and 8.C).

7.D Dominant Effects in Heterozygous Offspring of Chimeras or in Mice Heterozygous for a Spontaneous or Induced Mutation

7.D.1 Dominant morphological or developmental problems

Whether or not the ES cell chimeras demonstrate a phenotype, a dominant effect of the mutation could show up in the heterozygous offspring of the chimeras any time after fertilization. A

---

*MM3 refers to *Manipulating the mouse embryo: A laboratory manual*, 3rd edition (Nagy et al. [2003]. Cold Spring Harbor Laboratory Press, Cold Spring Harbor, New York).

dominant effect that causes the death of heterozygous embryos such that no heterozygous offspring are recovered from chimeras that transmit the ES cell coat color genotype (Section 3.C.3.b) can be handled in one of several ways. One solution is to analyze the phenotype directly in the offspring of the chimeras using the diagnostic procedures outlined in Chapter 5 for recessive mutations. If you have a good supply of chimeras that transmit approximately 100% of the ES cell genotype, then this might be feasible, although eventually you will be faced with replenishing the stock of chimeras to keep up the supply of embryos. Another alternative is to use a rescuing transgene (Box 7.1). Finally, you could consider producing a conditional allele to perpetuate the mutant allele in the germ line of mice but produce the mutant effect at a time of your choosing for further analysis (see Section 8.C). At the same time, you might also want to test-breed the chimeras with a different strain of mouse, just in case there is a strong genetic background effect and the mutants could survive.

Less severe developmental or physiological defects might also show up as dominant effects of the mutation in the offspring of the ES cell chimera or in mice heterozygous for spontaneous or induced mutations. For example, a skewing of the sex ratio in the heterozygous, ES cell-derived offspring of chimeras could indicate a dominant effect. If all heterozygous offspring are phenotypically female, you may have a dominant sex reversal effect (Section 3.C.1.c). However disconcerting this may be, the XX heterozygous females should breed just fine and the mutation can be established in the germ line. The dominant sex reversal phenotype can then be studied in the heterozygous XY offspring of these mice.

Other defects might be compatible with development to term or beyond, but still show up as dominant phenotypic effects (Figure 7.4). Be on the lookout for any characteristic phenotype associated with the heterozygous genotype and analyze it in the same way that you would a homozygous mutant phenotype (Chapter 6), with the caveat that phenotypes caused by imprinted genes must be approached differently (see Section 7.E). If it is a readily detectable abnormality and is 100% penetrant in heterozygotes, you can even use the dominant trait to genotype the animals. Do not make the assumption, however, that you have a dominant, as opposed to semidominant, mutation. It may be that the homozygous mutant phenotype is even more severe (Figure 7.4) and can provide further information regarding gene function. Thus, you will want to mate heterozygotes together to produce litters with all three genotypes for a complete assessment of the heterozygous and homozygous mutant phenotypes.

7.D.2 Dominant reproductive problems

Even if you had no trouble getting germ-line transmission from chimeras, you may not be home free from dominant reproductive problems. Dominant effects on the process of gametogenesis or other aspects of reproduction such as behavioral modifications in heterozygote males, females, or both sexes could cause sterility or subfertility (Sections 3.D.b and 3.D.c), and may only show up in the next generation. You will be able to study this effect in heterozygous mice as outlined in Chapter 6, but it will be more difficult to obtain the homozygous mutant animals, which could potentially have more severe or additional phenotypes. If only one sex is affected, for example, as in dominant sex reversal (Section 3.D.d), the mutation can be perpetuated by breeding from the unaffected sex, and then you can try assisted reproduction techniques for the affected sex (Section 4.D.3) when it comes time to produce homozygous mutants.

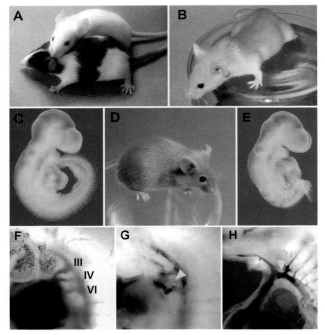

Figure 7.4. Examples of dominant or semidominant effects of mutations. (*A*) The gene *Kit*, formerly known as *Dominant White Spotting*, encodes a transmembrane receptor tyrosine kinase. Animals heterozygous for the allele, Kit^{W-v}, have large white spots (as well as other phenotypic effects), whereas homozygotes are black-eyed, all-white mice. (*B*) The *lethal yellow* allele of the *agouti* locus is embryonic lethal when homozygous but heterozygotes are viable, with a yellow coat. (*C–E*) The *Brachyury* or *T* locus encodes a transcription factor that affects mesoderm development. Heterozygotes for the T^{Wis} allele have no tails (*D*), whereas homozygous mutants have no posterior mesoderm beyond the lumbar region (*E*) and do not survive. A wild-type embryo (*C*) is shown for comparison. (*F–H*) The T-box transcription factor gene, *Tbx1*, has a number of phenotypic effects, including some that show a semidominant pattern of inheritance. India ink injection into the heart (*bottom left of each panel*) of midgestation embryos outlines the developing aortic arch arteries. Normal embryos (*F*) have three aortic arches (III, IV, and VI), whereas heterozygous $Tbx1^{tm1Pa}$ embryos (*G*) have a hypoplastic IVth arch artery (*arrowhead*) and homozygous mutant embryos (*H*) have only a single aortic arch artery (*arrowhead*). (*C,E* Courtesy of Debbie Chapman; *F–H* reprinted, with permission, from Jerome and Papaioannou, *Nat. Genet. 27:* 286–291 [2001 ©Nature Publishing Group].)

7.E Special Considerations for Imprinted Genes

By definition, an imprinted gene will have different activity depending on the parent of origin. Thus, if a mutation in a paternally inactivated imprinted gene is inherited from the father, it has no effect. However, if that same mutation is inherited from the mother, it will be the only active allele and the mutant effect will appear to be dominant. Once you have established that you have an imprinted gene by demonstrating that the mutant effect is present in heterozygotes only when the mutant allele is inherited from one parent but not the other (Figure 7.5A), you can study the mutant phenotype by breeding unaffected heterozygotes to wild-type mice to produce affected heterozygotes and then follow the procedures in Chapters 5 and 6 to characterize the

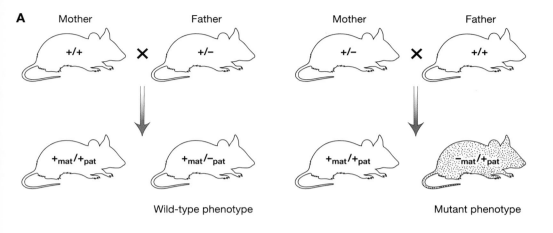

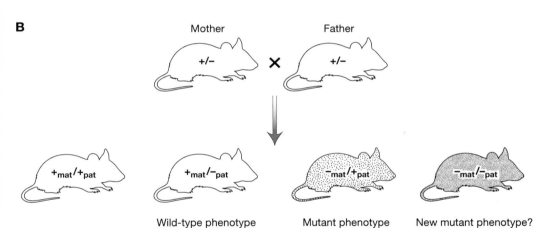

Figure 7.5. Breeding scheme for imprinted genes. (*A*) In this example of a paternally imprinted (inactivated) gene, reciprocal crosses establish that the heterozygous (+/−) phenotype is different depending on whether the mutant allele is inherited from the mother or father. To study this dominant effect, heterozygous females are crossed with wild-type (+/+) males to obtain heterozygotes with the mutant phenotype. (*B*) Heterozygotes can also be bred together to produce offspring of all possible genotype/phenotype combinations. If imprinting is ubiquitous, it is expected that the affected heterozygotes will have the same phenotype as homozygous mutants. However, if imprinting is tissue specific such that some tissues escape imprinting, homozygous mutants could have a more severe mutant phenotype.

HELPFUL HINT

To make the most of your hard-won mouse reagents, a targeted mutation in an imprinted gene can be used to provide information about the tissue specificity of imprinting by detecting allele-specific expression patterns. In a cross in which the wild-type allele is inactivated, simply use a cDNA probe for a genomic region deleted by a targeted mutation to see whether any tissues express the wild-type allele. If so, it would be indicative of tissue-specific escape from imprinting in that tissue.

phenotype. However, you should also breed unaffected heterozygotes together to produce homozygous mutants because they may have a different or more severe phenotype, which could come about if imprinting is not complete or if it is tissue specific. In this cross, the litters will contain wild-type controls, both affected and unaffected heterozygotes, and homozygous mutants, providing all possible genotype/phenotype combinations for study (Figure 7.5B).

CHAPTER 8

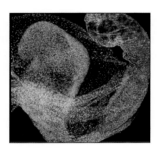

Getting Around an Early Lethal Phenotype

EMBRYONIC OR FETAL LETHALITY MAY BE THE LAST THING you want or expect, and you may be looking for a means to circumvent early lethality so that later stages of development, specific tissues, or even the adult mouse can be studied. The nature of development is such that the earliest effect of a mutation can potentially affect all subsequent stages, but the earliest time of a mutant effect may not correspond to the stage that you wanted to study or it may not be the only stage at which the gene is expressed.

This chapter deals with ways to get around an early lethal phenotype or even a tissue-specific detrimental effect that may not be lethal, so that you can study later time points without worrying about whether the later effects are primary or secondary. For example, a gene may be expressed throughout the development of a particular organ, but if a mutation prevents the organ from initially forming, the role of the gene in the later development of the organ cannot be studied. We have already mentioned the possibilities offered by cell, tissue, or organ culture to study specific mutant effects (see Section 5.E.2.d for an overview of the types of tissues that are amenable to study this way) that can essentially get around an embryo-lethal phenotype. In addition, assisted reproduction techniques (Section 4.D.3) can sometimes circumvent detrimental effects on fertility.

In this chapter, we present other ways of getting around an early phenotype (lethal or not). One or more of these methods may be useful for the analysis of any gene that has a complex temporal and spatial pattern of expression:

1. *The use of chimeras* (Section 8.A) can circumvent some lethal effects that are tissue specific and can also provide a tool for analysis of mutant effects.

Chapter opening artwork provided by Guy Eakin, Kat Hadjantonakis, Virginia Papaioannou, and Richard Behringer.

2. *Transgene rescue* (Section 8.B) can be used to supply a missing protein to allow further development of a mutant animal.

3. *Conditional alleles* (Section 8.C) allow you to confine the mutant effect to specific times, cell types, or cell lineages, effectively circumventing a mutant problem in some part of the embryo to reveal a mutant effect in others.

4. *Creating regulatory mutations* (Section 8.D) is another way of producing tissue-specific ablation or quantitative reduction of a gene product.

8.A Use of Chimeras

Combining mutant cells with wild-type cells in a chimera is a means of circumventing certain embryonic lethal effects. Combining embryonic cells from two different embryos (or embryonic cells with stem cells) in a manner that will allow them to co-mingle and develop as a single organism (Figure 8.1), commonly known as a chimera, is a time-honored embryology trick that was first used successfully in mammals nearly half a century ago. The basic idea is that the developmental potential of the two contributing cell types is revealed by their behavior in combination in a single developing embryo. The method has been used to trace cell lineages, explore the range of developmental potential or plasticity of different cells, and test cells for totipotency. It is also a valuable method for the analysis of a mutant phenotype.

8.A.1 What chimeras reveal

For our purposes, chimeras can be used either to explore the potential of cells carrying a mutation in your favorite gene (Section 5.D.2.d) or for the practical purpose of circumventing lethality if viability is limited by the failure of a critical tissue early in development. With chimeric "rescue," you essentially combine mutant cells with wild-type cells and "ask" the wild-type cells to make up for the critical function missing in the mutant cells. If rescue works, you may bypass an early lethality, allowing the chimera to develop far enough for you to study later stages at which the gene might have additional roles. For exploring the potential of mutant cells, the behavior of cells in a chimera can tell you something about the nature of the mutant effect. For example, the mutant cells may function normally in the presence of wild-type cells if the wild-type cells supply a factor missing from the mutant cells. This would indicate a non-cell-autonomous effect of the mutation. Conversely, a cell-autonomous defect would still affect mutant cells whether or not they were developing in the presence of wild-type cells in a chimera.

A common outcome of chimera experiments is that the distribution of mutant and wild-type cells is nonrandom or nonuniform when compared with chimeras made by the same method with two wild-type embryos (or a wild-type embryo and wild-type stem cells) (Figure 8.2). Depending on where and when the gene is normally expressed and the nature of the nonrandom distribution, this might represent useful information bearing on the mutant effect. Exclusion of mutant cells from an organ or tissue indicates that the mutant cells, for some reason, cannot compete with the wild-type cells in the development of that organ or tissue.

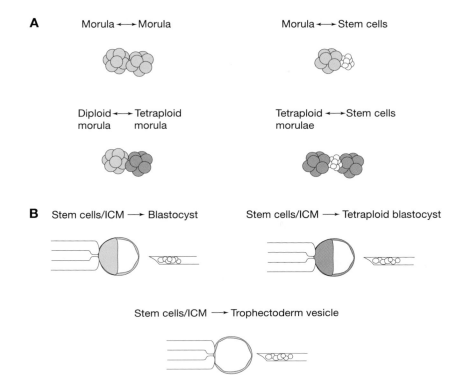

Figure 8.1. Different ways of making chimeras. (*A*) Aggregation at the morula stage. Embryos at the morula stage (2.5 dpc) can be aggregated together or aggregated with stem cells (either embryonic stem or trophectoderm stem cells). Tetraploid morulae can be used in combination with a diploid morula or stem cells to produce tissue-restricted chimeras. In the case of aggregation of tetraploid morulae with stem cells, two morulae are used to increase the chance of successful development. (*B*) Injection of cells into the cavity of 3.5-dpc blastocysts. Isolated inner cell mass (ICM) cells, whole ICMs, or stem cells can be injected into diploid or tetraploid blastocysts. In addition, an ICM can be injected into an isolated trophectoderm vesicle. All injection methods produce tissue-restricted chimeras. See Box 8.3 for the tissue composition of resulting chimeras.

However, there are many ways to arrive at a nonrandom mix of cells in a chimera, and the challenge is to figure out which applies. For example:

- The mutant cells may die long before the chimera reaches a particular point in development.
- Mutant cells may have a lower proliferative rate than normal and are thus overgrown by wild-type cells.
- Mutant cells may not be able to respond to/emit normal inductive signals.
- Mutant cells may be inhibited in or incapable of cell movement or migration necessary to form an organ or tissue.
- Surface properties of mutant cells may render them incapable of interacting with wild-type cells.
- Mutant cells may be unable to complete a specific step in differentiation.

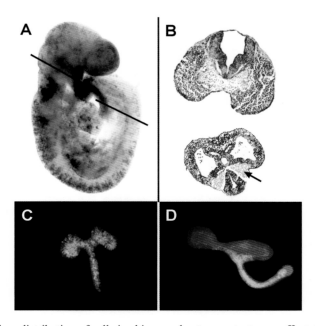

Figure 8.2. Nonrandom distribution of cells in chimeras due to a mutant gene effect. (*A*) Whole mount of a chimera made by blastocyst injection of *Tbx6* homozygous mutant ES cells carrying a ubiquitously expressed *lacZ* transgene (*ROSA26*) as a cell marker. The β-gal-positive mutant cells (*blue*) are distributed throughout the embryo but are much less dense in the posterior region, except for the tail bud, where the gene is normally expressed, which is composed mostly of mutant cells. (*B*) Section of a similar chimera counterstained with eosin. The lines in *A* represent the approximate level of the section. This chimera has a high level of contribution of mutant ES cells to all tissues with the notable exception of the somites, which are composed almost entirely of wild-type cells (*pink, arrow*). The mutation causes mutant cells to accumulate in the tail bud and prevents them from taking part in somite formation. (*C,D*) Kidney primordia from chimeras made by injection of ES cells carrying a *Hoxb7/GFP* transgene into blastocysts. The ureteric bud epithelium was stained with anticytokeratin (*red*). The chimera in *C* was made with green fluorescent protein (GFP)-labeled, wild-type ES cells and shows a random contribution of GFP-positive cells to the ureteric bud. The chimera in *D* was made with GFP-labeled ES cells homozygous mutant for the Ret receptor tyrosine kinase. The mutant ES cells contribute only to the ureter and trunks of the ureteric bud branches but are excluded from the tips of the ureteric buds. (*A* Courtesy of Debbie Chapman; *B* reprinted, with permission, from Chapman et al., *Mech. Dev. 120:* 837–847 [©2003 Elsevier]; *C,D* courtesy of Reena Shakya and Frank Costantini.)

Any combination of altered properties could lead to segregation or sorting of cells in chimeras. It is necessary to examine a number of chimeras at different stages of development to narrow down these options. Better yet, visualization of the mutant cells in living chimeric embryos or tissue explants might be possible using short-term embryo culture and vital fluorescent markers to distinguish the mutant from wild-type cells (Figures 8.2C,D and 8.3B–F, and see Figure 5.6). Of course, a prerequisite for making use of any chimeric combination is that you have some means of distinguishing between the two component cell types at the time of analysis (see Box 8.1). Similarly, it is unlikely that you will be able to distinguish the mutant

Figure 8.3. Distinguishing cells in a chimera using *GFP* transgenes. (*A*) Detection of chimerism and differentiation in the allantois of an 8.5-dpc embryo using immunohistochemistry. This chimera was made with a wild-type embryo and a transgenic embryo carrying a ubiquitous, nuclear-localized *GFP* transgene. The embryo was subjected to sequential immunohistochemical labeling with anti-GFP and anti-PECAM antibodies with alkaline phosphatase- and peroxidase-conjugated secondary antibodies, respectively, sectioned, and counterstained with Nuclear Fast Red. Anti-GFP immunohistochemistry and the counterstain distinguish cells as derived from either the wild-type embryo (*pink nuclei*) or *GFP* transgenic embryo (*brown nuclei, arrows*). Additionally, immunohistochemistry for PECAM, a molecular marker of endothelial differentiation, identifies cells that have begun to form the endothelium of blood vessels (*blue cytoplasmic label, arrowhead*), regardless of their origin. (*B*) Display of both chimeric components in an adult heart using two different fluorescent transgene markers. The chimera was made by aggregating a morula carrying an enhanced cyan fluorescent protein (ECFP) transgene with a morula carrying an enhanced yellow fluorescent protein (EYFP) transgene. The image is a double exposure using ECFP and EYFP filters, consecutively.

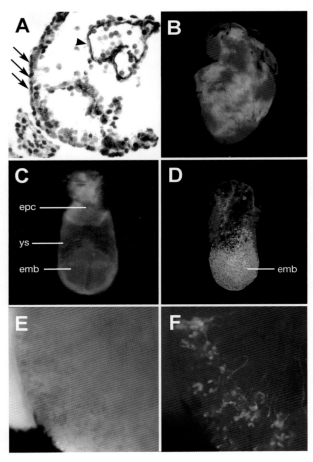

(*C,D*) Comparison of cytoplasmic and nuclear-localized fluorescent markers in whole-mount gastrulation-stage chimeric embryos. (*C*) A 7.5-dpc chimera made by aggregation of a tetraploid EYFP transgenic morula and a diploid ECFP transgenic morula. In both cases, the fluorescent protein is cytoplasmically localized. The diploid cells make up the entire embryonic region (emb) and the yolk sac (ys) mesoderm, whereas the tetraploid cells are confined to the extraembryonic tissues including the ectoplacental cone (epc) and yolk sac endoderm. Image is a double exposure using ECFP and EYFP filters, consecutively. (*D*) A 6.5-dpc chimera made by injecting ES cells transgenic for a nuclear-localized EGFP into a wild-type blastocyst. The chimeric embryo has been counterstained with FM 4-64 (Molecular Probes, Oregon) to reveal cell boundaries (*red*). The progeny of the ES cells are in the embryonic region (emb). (*E,F*) Bright-field (*E*) and fluorescence (*F*) images of a 400-μm vibratome section of an adult kidney from a chimera made with wild-type and ECFP transgenic cells. The fluorescent cells are in the minority, making the structural details easy to see. (*A* Courtesy of L.A. Naiche and Virginia Papaioannou; *B* and *C* reprinted from Hadjantonakis et al., *BMC Biotechnol.* 2: 11 [©2002 licensee BioMed Central Ltd.]; *D* courtesy of Guy Eakin, Kat Hadjantonakis, Virginia Papaioannou, and Richard Behringer; *E,F* courtesy of Kat Hadjantonakis.)

embryos from wild type at the time you make the chimeras, so you will need to have a means of genotyping the chimeras at the time of analysis that will distinguish heterozygous from homozygous mutant cells (see Box 8.2).

Box 8.1 Distinguishing Cells in Chimeras

Identifying the different cells making up a chimera is a simple matter of having a marker in one or the other cell components and detecting the marker at the time of analysis. Ideally, a marker should be stable, heritable, cell localized, cell autonomous, and developmentally neutral. It is also quite useful if the marker is ubiquitously expressed, easily detectable, and if several different variants are available.

Depending on the circumstances, a trade-off can sometimes be made with a less-than-ideal marker. One of the most common markers for detecting chimerism is coat color (see Figure 3.3), but the genes affecting coat color meet only a few of the criteria of an ideal marker. The pigmentation pattern is stable, heritable, and easily detectable with several variants. However, melanin granules are not always cell localized (they are secreted into the hair shaft) and some pigment genes affect coat color in a non-cell-autonomous way (e.g., agouti is a secreted signaling molecule). Also, pigment is certainly not ubiquitous. Nonetheless, if all you need to know is a rough approximation of the level of chimerism, as in assessing embryonic stem (ES) cell chimeras, this imperfect marker serves the purpose, and this noninvasive means of detection leaves the animal intact for other purposes.

On the other hand, when investigating the behavior of mutant cells in combination with wild-type cells in a chimera, the most useful marker is one that allows positive identification of the origin of each and every cell either in histological sections, whole mounts, or tissue explants. It is also useful to be able to combine the assay for the chimera marker with assays for the molecular differentiation of tissues (Figure 8.3A). Although there are some intrinsic markers, such as isozyme variants of housekeeping genes that can be detected histochemically, detection may require tissue homogenization, thus sacrificing spatial information. The most commonly used markers are introduced transgenes that have been demonstrated to be developmentally neutral and are more or less ubiquitously expressed. These transgenes are heritable, stable, and selected for their cell-autonomous, cell-localized properties. The Gt(ROSA)26Sor gene trap insertion (commonly called *ROSA26*) is perhaps the most commonly used marker (see Figure 8.2A,B). This is a gene trap insertion into a gene that is ubiquitously expressed in the embryo (but not in the extraembryonic tissues) and can be easily detected by X-gal staining (MM3).* It appears to have no significant detrimental developmental effects even when homozygous, and it is thus a simple matter to breed the transgene onto the mutant stock of mice before making chimeras.

Increasingly, fluorescent protein genes are being fashioned into transgenic markers. For chimera studies, a transgenic, ubiquitously expressed, developmentally neutral, nuclear-localized green fluorescent protein (GFP) comes close to providing the ideal marker to identify every cell in a chimera (Figures 8.2 and 8.3). Fluorescent proteins have the advantages of the availability of several distinguishable spectral variants (Figure 8.3B,C), the possibility of creating fusion proteins to localize the marker to different subcellular compartments (Figure 8.3C,D), and the feasibility of detection in living or fixed tissues either by fluorescence microscopy or by immunolocalization (Figure 8.3A). In addition, embryos can be imaged vitally with time-lapse microscopy for a dynamic picture of development. GFP fusion genes can be targeted to specific loci, such as the *ROSA26* locus, or random transgene insertions can be selected for ubiquitous expression. Mice carrying such markers are commercially available. (For a list of useful markers, see MM3.)

Either component of a chimera could theoretically be the marked component. However, because it is easier to detect a few positive cells in a background of negative cells than vice versa, whenever possible the marked cell population should be in the minority. This will usually be the mutant cell population, so plan accordingly.

---

* MM3 refers to *Manipulating the mouse embryo: A laboratory manual*, 3rd edition (Nagy et al. [2003], Cold Spring Harbor Laboratory Press, Cold Spring Harbor, New York).

Box 8.2 Genotyping Chimeras

If you are making chimeras between wild-type embryos and homozygous mutant ES cells, there will be no need to genotype the chimeras because they will all be +/+ ↔ –/–. If the mutant ES cells also contain an independent marker (see Box 8.1), detecting the marked cells is the same as detecting the mutant cells. However, if you are making chimeras with embryos rather than ES cells, you will necessarily be making chimeras from +/+, +/–, and –/– embryos if you are breeding from heterozygotes because the mutation is lethal. Furthermore, the usual polymerase chain reaction (PCR) genotyping protocols that depend on detection of a mutant and wild-type allele will not help, because +/+ ↔ –/– chimeras have both alleles and look the same on a PCR gel as +/+ ↔ +/– chimeras. There are several possibilities for solving this problem:

1. *Use two different mutant alleles that you can distinguish by PCR* (Figure 8.4). This is not as difficult as it sounds because you will often have two alleles that act the same, i.e., alleles before and after excision of a selection cassette that both act as null alleles, or alleles with two different selectable markers. Detection of both null alleles ensures that the chimera is +/+ ↔ –/–.

2. *Make use of tissue restriction in chimeric combinations to provide "pure" tissue for genotyping.* For example, if you inject wild-type inner cell masses (ICMs) into blastocysts from heterozygous mutant matings, the trophectoderm derivatives and primitive endoderm will be derived from the blastocyst (see Box 8.3). Thus, these tissues can be isolated from the chimeras for PCR genotyping without fear of contamination from the +/+ ICM derivatives. However, beware of maternal contamination when dissecting trophectoderm derivatives for PCR.

3. *Use laser capture microdissection to isolate a sample composed of pure marked or unmarked cells*, depending on which component has the marker, for PCR genotyping.

4. *Use statistics.* This is probably the least satisfactory method, because it requires a large number of chimeras to be analyzed and depends on the existence of a clear, consistent difference in cell behavior between +/+ ↔ –/– chimeras and +/+ ↔ +/– or +/+ ↔ +/+ chimeras. There will always be doubt as to the interpretation but you may have no other choice. Be sure to analyze a sufficient number of chimeras for statistical significance (see Box 5.1, pp. 72–73).

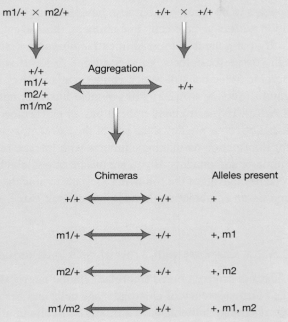

Figure 8.4. Genotyping chimeras using two different null alleles. Two different null alleles, m1 and m2, which can be distinguished by Southern or PCR, are used. In one cross, m1 heterozygotes are bred with m2 heterozygotes, yielding embryos with four unique genotypes. Wild-type embryos are generated in the second cross. Morulae from the first cross are aggregated with morulae from the second cross to generate chimeras. The four different genotypic combinations of chimeras that are generated can be identified by Southern or PCR analysis of tissues. The chimeras with homozygous mutant and wild-type cells will have all three alleles (m1, m2, and +). The other chimeras can serve as controls.

> **HELPFUL HINT**
>
> *Pigmentation differences are useful markers of chimerism in postnatal mice and can also be used to detect chimerism in embryos. The retinal pigmented epithelium, after it becomes pigmented at 10.5 days postcoitus (dpc), provides an indicator of chimerism between pigmented and nonpigmented components.*

8.A.2 Making chimeras

Two basic methods for making chimeras include aggregation of embryos or cells at the morula (8- to 16-cell) stage of development (MM3) and injection of cells into blastocyst-stage embryos (MM3) (see Figure 8.1). If you obtained your mutation through gene targeting, then you will be familiar with one or the other of these methods for making ES cell chimeras. Both methods require basic skills in embryo recovery, handling, and transfer, whereas the injection method also requires specialized microinjection equipment and skills. Another variable in making chimeras is the nature of the cells used. There are few choices, because only embryonic cells and stem cells from embryos are able to make chimeras with preimplantation embryos. However, both diploid and tetraploid cells can be used, and the combinations of genotype, ploidy, and method provide plenty of variety. Each combination potentially provides a different type of information because different types of cells have different developmental potential and/or some cells may offer a selective advantage over others (Box 8.3). Keep in mind that in addition to developmental limitations imposed on the two components of the chimeras by (1) cell type and (2) genotype with respect to the mutant allele, there may also be (3) genetic background effects that skew the composition toward one component or the other. It is also worth considering using heterozygous as well as wild-type cells as controls, even if the mutation appears to be recessive, because heterozygous cells may not fare well when placed in competition with wild-type cells in a chimera, thus revealing a heterozygous phenotype.

If your gene is expressed in extraembryonic tissues and/or you want to examine every possible tissue for defects, go to Section 8.A.2.a. If you are not interested in looking at extraembryonic tissue but still want a chimeric mix in all tissues of the fetus, you can do morula aggregation (Section 8.A.2.a) and you also have the option of using blastocyst injection (Section 8.A.2.c). If your mutant embryo has an extraembryonic defect and your goal is to rescue the lethal effect or pin down the site of the extraembryonic defect, go to Section 8.A.2.b or 8.A.2.c for information on making chimeras with lineage-restricted contributions of cells from the two chimeric components. If you are unsure about which kind of chimera will best suit your needs, read through all of the following sections to note the limitations and advantages of each method and tissue combination. In addition, consult Box 8.3.

8.A.2.a Chimeras with a mix of cells in all tissues

Cleavage-stage embryos (morulae) can be aggregated together to make chimeras. Because of the pluripotentency of cleavage-stage cells and because cell mixing is unimpeded at this time, the resulting chimeras have a mixture of cells in all tissues, both embryonic and extraembry-

Box 8.3 Developmental Limitations of Cells in Chimeras

When making chimeras, you can take advantage of natural limitations in the developmental potential of different components to construct a chimera with specific tissue composition. The following table shows possible fates of different kinds of cells in a chimera made at early embryonic stages by aggregation or at a slightly later stage by blastocyst injection. Tissue is diploid unless otherwise stated.

Tissue used to make the chimera	Fate of tissue in the chimera at a later stage of development			
	Trophectoderm and its placental derivatives	Yolk sac endoderm and parietal endoderm	Yolk sac mesoderm and other extra-embryonic mesoderm	Embryo proper
Morula	✓	✓	✓	✓
Blastocyst	✓	✓	✓	✓
ICM		✓	✓	✓
ES cells			✓	✓
Tetraploid morula	✓	✓		
Tetraploid blastocyst	✓	✓		
Trophectoderm stem (TS) cells	✓			
Trophectoderm vesicle	✓			

Because of this limitation in tissue developmental potential, chimeras with different compositions of tissue can be made by choosing different tissue combinations and using either morula aggregation or blastocyst injection as shown below. As above, tissue is diploid unless otherwise stated.

Method	Components of chimera	Origin of trophectoderm in chimera	Origin of yolk sac and parietal endoderm in chimera	Origin of fetus and extra-embryonic mesoderm in chimera
Aggregation	morula ↔ morula	mixed	mixed	mixed
	morula ↔ ES cells	morula	morula	mixed
	morula ↔ TS cells	mixed	morula	morula
	morula ↔ tetraploid morula	mixed	mixed	morula
	ES cells ↔ tetraploid morula	tetraploid morula	tetraploid morula	ES cells
Injection	ICM → blastocyst	blastocyst	mixed	mixed
	ES → blastocyst	blastocyst	blastocyst	mixed
	TS → blastocyst	mixed	blastocyst	blastocyst
	ICM → tetraploid blastocyst	tetraploid blastocyst	mixed	ICM
	ES → tetraploid blastocyst	tetraploid blastocyst	tetraploid blastocyst	ES cells
	ICM → trophoblast vesicle	trophoblast vesicle	ICM	ICM

onic. Extraembryonic tissue refers to the fetal membranes, trophoblast, chorion, yolk sac, amnion, and allantois, in contrast to the embryonic tissue, which is the fetus or embryo proper. In fact, aggregation of morulae is the only way to get chimeras with a mix of the two components in every tissue. If one component of the chimera is mutant, however, the composition of the chimeric mix may be affected by alterations in the developmental potential of the mutant cells, and this, of course, is what you will be looking for in the analysis of the chimeras. The production of aggregation chimeras is straightforward (MM3) and consists of isolating embryos, removing the zona pellucida, culturing the embryos in contact with one another until they make a single aggregate, and then transferring the aggregates to a foster mother. Because in most cases the mutant genotype will not be evident at the time of aggregation, all of the embryos resulting from heterozygous matings are aggregated with wild-type embryos and the genotypes of the chimeras are sorted out later during the analysis (see Box 8.2). Chimeras made with heterozygous or wild-type cells will serve as useful controls for technique and for possible genetic background effects. In addition to being able to figure out the genotype of the embryos, the two components must also be distinguishable through the use of a cellular marker to complete the analysis of the cellular distribution within the chimera. The resolution of the analysis will only be as good as the resolution of the cellular marker used, so a stable, cell-based, developmentally neutral marker is essential (see Box 8.1).

HELPFUL HINT

If you want to perform a chimera experiment using an existing homozygous mutant ES cell line, it is easiest to mark the recipient morula or blastocyst (e.g., with ROSA26*). However, this may not be optimal because it is usually best to mark the mutant cells of the chimera for easier detection of their fate. What can you do? One possibility is to target the* ROSA26 *locus or another ubiquitously expressed locus of the mutant ES cells with a knock-in marker construct such as* lacZ, *alkaline phosphatase, or a fluorescent protein. One technical consideration is to use a selectable marker for the targeting that is not already present in the ES cell line. For example, you cannot use G418 selection if the ES cell line is already* neo *resistant.*

The analysis of chimeras should begin around the time of lethality of the homozygous mutant. A thorough morphological assessment at this and later stages, similar to that done in determining the nature of the mutant effect in the first place (Chapter 5), will indicate whether or not the phenotype is altered by the chimeric combination. Remember that there will be some stochastic variation in the contribution of the different components to the chimeric tissues, independent of the genotypes, so you can expect to see a range in the overall percentage contribution of mutant and wild-type cells. A range in chimeric composition is useful because it allows you to study the behavior of mutant cells in combination with different proportions of wild-type cells. The results may look different when the mutant cells are in a majority or a minority. Although chimeras with a high proportion of mutant cells, so-called high-level chimeras, might look very similar to the homozygous mutants, low-level chimeras might have a mild phenotype and survive past the time of mutant lethality—in other words, a partial rescue. This will allow the examination of mutant cells in later-stage embryos. The type of analysis that you do will be similar to those that investigate the mutant alone, but in the case of chimeras, you will be looking at the phenotype in relation to the quantity and location of mutant cells.

If the phenotype of the mutant is an early lethality caused by a defect in extraembryonic tissues rescued in chimeras, you will be able to study the role of the gene in embryonic tissues at later stages in chimeras. You might also want to consider making lineage-restricted chimeras (Section 8.A.2.b) that will rescue the extraembryonic phenotype and allow you to control the chimeric composition in different tissues of the chimera, for example, by producing a fetus derived exclusively of mutant cells with a mixture of mutant and wild-type cells in the placenta.

8.A.2.b Lineage-restricted chimeras: Morula aggregation with stem cells or tetraploid embryos

1. *Morula ↔ ES cells.* By using the relatively simple method of morula aggregation to make chimeras you can exploit the developmental restrictions of certain cell types to produce chimeras that are not random mixes of cells. If you have homozygous mutant ES cells, you can aggregate the ES cells with normal, wild-type morulae. (Incidentally, this saves you the trouble of genotyping the chimeras because they are made with ES cells of known genotype.) The resulting chimera will be a complete mix of components in the embryo proper and the mesoderm of the extraembryonic tissues. However, because of inherent restrictions in ES cell developmental potential, the trophectoderm, its placental derivatives, and the extraembryonic endoderm will all be derived solely from the morula (see Box 8.3). This combination might be useful if the mutant phenotype you have is limited to the placenta or yolk sac endoderm, in which case the mutant phenotype would be rescued in the chimeras. It can also help to distinguish a yolk sac endoderm defect from a yolk sac mesoderm defect by providing a yolk sac with the endoderm made up entirely of wild-type cells and the mesoderm made up of a mixture of wild-type and homozygous mutant cells. (If you need to distinguish the affected layer for a yolk sac phenotype, mutant and wild-type cells can be completely separated into one layer or the other using tetraploid embryos and ES cells; see below and MM3.)

2. *Morula ↔ TS cells.* If TS cells are available, they can be used to make lineage-restricted chimeras. In this case, the derivatives of the TS cells are restricted to the trophectoderm, whereas the morula will contribute to the trophectoderm as well as make up the entire embryo proper, yolk sac, and parietal endoderm (see Box 8.3). A chimeric composition using wild-type TS cells and mutant morulae might be useful to rescue a trophectoderm defect if the mutant phenotype is limited to the trophectoderm in the early stages.

3. *Diploid morula ↔ tetraploid morula.* Another means of stacking the odds to control the composition of chimeras is to use tetraploid embryos. Tetraploid embryos can be made by electrofusion of the two blastomeres of a two-cell embryo (MM3). By themselves, tetraploid embryos develop very poorly, and aggregated with diploid morulae, their cells are at a competitive disadvantage and only contribute well to certain cell lineages (see Box 8.3). In the context of mutant analysis, this method is sometimes called tetraploid rescue because it can be used to circumvent lethality caused by a defect in the trophectoderm, the placenta, or the yolk sac endoderm. If you know that this is the cause of lethality in your mutant and you want to study the role of the gene in other tissues in the later fetus, or even in adults, this method could provide a means to do so.

In a chimera, although they may be at a disadvantage, tetraploid cells are capable of making the trophectoderm of the blastocyst and its derivatives, as well as the primitive endoderm and its derivative, the yolk sac endoderm, but they are not so effective when it comes to making the ICM and its derivatives. That means that in a tetraploid ↔ diploid chimera, the tetraploid cells are at a severe selective disadvantage compared with the diploid cells in contributing to the ICM, but they can form a chimeric trophectoderm or yolk sac endoderm and possibly rescue a mutant effect in these tissues. A chimera made by aggregating a tetraploid embryo with a diploid embryo will develop with a mixed population of cells in the trophectoderm and primitive endoderm but with primarily diploid cells in the ICM derivatives (the fetus and extraembryonic mesoderm). However, do not forget that the chimera starts out with a mix of diploid and tetraploid cells, and the cells sort out into separate compartments only gradually. This selection likely occurs between the blastocyst stage and gastrulation, but the precise timing is not known and could affect interpretation of chimeric patterns depending on the time of analysis and when your mutation acts.

4. *Diploid ES cell ↔ tetraploid morula.* For an even more extreme separation of chimeric components, homozygous mutant ES cells can be used as the diploid component of the chimera. ES cells have essentially the same reciprocal developmental potential compared with tetraploid cells in chimeras: They are capable of contributing to all ICM derivatives, but not to the trophectoderm or primitive endoderm lineages, so the resulting chimeras will have the components completely segregated (see Box 8.3). The nonmutant, tetraploid component will rescue any trophectoderm or endoderm defect, leaving the mutant ES-derived fetus to develop further. Its continuing development will thus allow you to search for developmental problems associated with the mutations in later-stage embryos more than previously possible. One complication is that an entirely ES-derived fetus will only survive to late gestation and birth if the ES cells are fully competent. Thus, the strain and passage number of the ES cells may be important for later development.

If your mutation results in a yolk sac defect, ES cell ↔ tetraploid chimeras might also be an effective way to sort out whether the defect has its origin in the yolk sac endoderm or the yolk sac mesoderm, because this method results in a yolk sac with the two components restricted to one layer or the other of this bilaminar structure. The reciprocal combination of a mutant tetraploid embryo and wild-type ES cells provides the opposite distribution of mutant and wild-type cells in the yolk sac.

The disadvantages of these methods are that specialized equipment is required to make tetraploid embryos, and it may be an effort to derive mutant ES cells. However, if you think that there is a good chance that a wild-type trophectoderm and/or endoderm layer will get around the lethal problem or give you valuable information about the tissue restriction of a mutant effect, it may be worth the effort.

8.A.2.c Lineage-restricted chimeras: Blastocyst injection

Chimeras made using blastocyst injections (MM3) are all unbalanced because the trophectoderm and ICM have already been determined by the blastocyst stage and all cells that are successful at making chimeras at this stage have developmental restrictions. The ICM has the least restricted potential and can contribute to the fetus as well as both layers of the yolk sac in a chimera: ES cells can contribute to the fetus and extraembryonic mesoderm, but not the extraembryonic endoderm; TS cells are limited to contributing to the trophectoderm (although contribution levels are very low); and tetraploid cells are limited to trophectoderm derivatives and extraembryonic endoderm (see Box 8.3). If you are interested in examining mutant cell behavior within the fetus and/or in the yolk sac and fetal membranes, but not the placenta, you can isolate ICMs from embryos of heterozygous matings and inject them into wild-type blastocysts. If you have homozygous mutant ES cells, you can inject these into wild-type blastocysts, and the only additional limitation will be that the yolk sac endoderm will not be chimeric. If your goal is to rescue a trophectoderm or yolk sac endoderm defect to let mutant embryos survive longer for study of later stages, you might try injecting mutant ICM cells or mutant ES cells into tetraploid blastocysts. These combinations give a completely mutant fetus in a wild-type trophectoderm with either a mixed or tetraploid extraembryonic endoderm, respectively.

For the sake of completeness, we should mention that a completely polarized chimera can be made by injecting an ICM into a mechanically dissected trophectoderm vesicle, although the skills necessary for this procedure make it impractical for most purposes.

The analysis of any chimeric combination should begin around the time that a phenotype is first observed in the homozygous mutant embryos and continue as outlined in Chapter 5 if a rescue or partial rescue is achieved. Even with no "rescue," the behavior of the mutant cells in combination with wild-type cells in the embryo provides valuable information about their developmental potential. Before you start, have a look at Box 8.4.

Box 8.4 Checklist for a Chimera Experiment

Answers to the following questions will facilitate the best design for your chimera experiment. Some of the questions are based on the supposition that you know the identity of the gene that is mutated, though this information is not essential for performing a chimera study.

- Which tissue(s) expresses the gene product in question?
- At what stage does expression initiate in the tissue?
- When does the tissue form during embryogenesis?
- When does the mutant phenotype first become apparent?
- Which chimera system will you use (morula aggregation, blastocyst injection, ES cells, TS cells, tetraploid embryos, etc.)?
- What type of chimera will be your control? Usually, +/+ ↔ +/+ but potentially also +/– ↔ +/+.
- Have you controlled for genetic background differences?
- How will you genotype the chimeras?
- Which marking system will you use to follow the fate of mutant and/or wild-type cells (i.e., isozyme variant, pigmentation, *lacZ*, fluorescent protein, etc.)?
- How many chimeras will you need for the study?
- What range of chimerism will be useful to address your specific question?
- Which stage(s) and tissue(s) will you analyze?
- What types of analyses will you perform (gross anatomy, histochemistry, fluorescent imaging, histology, molecular markers)?

8.B Transgene Rescue

It is possible to bypass an embryonic lethality caused by a mutation by providing expression from a transgene to critical tissues (see Box 7.1 [p. 168] and Figure 7.3), where a transgene was used to rescue a dominant embryo-lethal phenotype. Transgene expression can be supplied by a heterologous regulatory element or endogenous regulatory elements. Because transgene expression is influenced by the site of integration, each transgenic mouse line will potentially provide a different extent of rescue of the mutant phenotype, leading to the equivalent of an allelic series of phenotypes. On a wild-type background, the expression of the rescuing transgene may result in a gain-of-function phenotype that could also be informative.

Generate transgenic mice on a wild-type genetic background with your rescuing gene construct and then backcross the transgene onto the mutant background to obtain homozygous mutant transgenic mice for analysis. If you are generating bacterial artificial chromosome (BAC) transgenic mice, you can use the entire BAC, allowing you to use the plasmid backbone as a tag to detect the presence of the BAC in mice from this cross. This does not ensure that the entire BAC has integrated intact but it is a quick and simple way to detect the presence of the foreign DNA. Alternatively, if the inbred strain from which the BAC is derived differs from the strain of the zygotes that you are using to make the transgenic, then you can use DNA polymorphisms between the strains to detect the presence of the BAC. You will find that genotyping a homozygous mutant with a transgene is much simpler in the case of a BAC transgene if you have two different mutant alleles (e.g., *neo* and *lacZ* knockout alleles). If you only have one mutant allele, you will have to use a DNA polymorphism to track the BAC on the mutant genetic background at the same time as genotyping for the mutant allele. This is because the genotyping strategies for the wild-type allele will yield the same result if you have a heterozygous or homozygous mutant along with the wild-type allele on the BAC. Of course, if you are using a BAC derived from a species other than the mouse, you can use species-specific probes or primers to detect the presence of the BAC in a mouse. When designing a transgene rescue experiment, carefully think through how you will follow transgene expression (northern, reverse transcriptase [RT]-PCR, in situ hybridization, etc.) on the wild-type and mutant genetic backgrounds.

There are a few caveats to the transgene rescue approach. First, it may be difficult to generate transgenic mice carrying a rescuing gene construct if overexpression of the gene is detrimental. If so, you may find that you only recover lines that express the transgene at low levels. This can be useful because these lines will probably lead to hypomorphic rescue of the mutant phenotype. If you find that the transgene is lethal on a wild-type genetic background, it would be possible, though not very practical, to inject the rescuing transgene construct directly into heterozygous and homozygous mutant zygotes from heterozygous crosses. In addition, if you are using a BAC transgene, do not forget that there may be other genes on the BAC that could influence the rescue and that expression of transgenes can be mosaic, which may influence the outcome of the experiment.

8.C Conditional Alleles

Making a conditional allele as described in Section 2.B.4 gives you the option of leaving gene function intact until the allele is exposed to the appropriate site-specific recombinase. Although a conditional allele with a positive selectable marker left in place may function as a wild-type allele, we recommend that you remove the selection cassette from the allele in ES cells before making mice (see Box 2.3, p. 17). This will circumvent any potential interference with gene function by the promoter of the selectable marker.

Once the conditional allele is in the germ line of mice, whether the selectable marker is left in or taken out, several things must be done to verify that the conditional allele is functioning as planned. First, you must ensure that the conditional allele actually functions like a wild-type allele and not a null or hypomorph. Of course, if it does function as a hypomorph (Section 4.E.2), it can still be a useful allele, and it can also be used conditionally to produce a tissue-specific null allele, provided that the mutant effects of the hypomorph are not too early or too severe. Second, you must show that the site-specific recombinase produces the desired effect, i.e., the removal of the floxed or flrted exons, resulting in a functionally null allele. Only then can you make use of the conditional allele for time- or tissue-specific ablation of gene function.

8.C.1 Testing a conditional allele

A conditional allele should have no effect on gene function. To prove this, one of the first things to do is to examine heterozygotes for the conditional allele and then breed the allele to homozygosity to ascertain that there is no mutant phenotype. In the case for which the null allele is recessive, you should also produce mice that are compound heterozygotes for the conditional allele and a null allele to show that the conditional allele is fully functional. If the homozygous and compound heterozygous mice are normal and fertile (determine statistics to be confident; see Box 5.1, pp. 72–73), you probably have a good conditional allele.

On the other hand, if a mutant phenotype is detected with the conditional allele, the most obvious culprit would be the positive selection cassette if it was left in the locus. Remove the cassette by breeding as outlined in Section 4.D.1 and then breed to homozygosity again. Keep the original allele, however, because it may prove to be a useful hypomorphic allele. If the selection cassette has already been removed, there is less chance of a mutant phenotype, but it is still possible if the remaining recombinase sites are located in regulatory regions. This type of allele must thus be tested for phenotypic effects as well.

If a conditional allele without a selection cassette has a mutant phenotype, you can still make use of it if it proves to be a hypomorphic rather than a null allele, because you will be able to ablate the function of this hypomorphic allele conditionally, provided it allows development of the tissue of interest. However, be aware that a hypomorphic mutation could well have an effect on the tissue of interest, so that when the null allele is produced by Cre recombinase, the development of the tissue might already be compromised.

The next step is to show that site-specific recombinases have the desired effect on the targeted conditional allele. This should have already been tested in ES cells (Section 2.B.4) but now that the allele is in mice, removing the floxed or flrted exon(s) serves the dual purpose of proving the utility of the allele and producing a null mutation. Breed the mice carrying the conditional allele with deleter mice as outlined in Section 4.D.1 and recover mice with the null allele. In addition to analyzing these mice for the effects of a null mutation (Chapters 5 and 6), you can also use them in the breeding scheme to produce a tissue-specific mutation (see Section 8.C.2), and also to test whether the conditional allele is wild type or hypomorphic. First, go to Section 4.E.2 and follow the breeding scheme to test for a hypomorphic or null allele. Then, proceed using the conditional allele.

8.C.2 Using a conditional allele

Once tested and proven to have no mutant phenotype on its own and to produce a null allele upon exposure to recombinase, a conditional allele can be used to ablate gene function either in specific tissues and/or at specific times by crossing mice carrying the conditional allele with mice carrying a *cre* (or other appropriate recombinase) transgene. Either tissue- or stage-specific ablation approaches can be used to allow an embryo to develop beyond an early lethal effect caused by a null mutation. Additionally, a tissue-specific ablation using a *cre/flp* transgene expressed only in that tissue is useful if you are mainly interested in a certain organ or tissue and want to distinguish primary from secondary effects. A ubiquitous, stage-specific ablation approach, accomplished by using an inducible, ubiquitously expressed Cre/Flp, allows you to study the role of the gene in multiple tissues later in development and could reveal later time points in development in which the lack of gene function causes lethality. For both approaches, the limiting factors, apart from space to breed the animals, are the number and variety of different recombinase transgenic mouse lines available and your ability to verify the excision event in specific cells or tissues.

There is an ever-increasing number of recombinase-expressing transgenic mouse lines available (see Appendix 1), but use caution in selecting one to make sure that it has been fully characterized with respect to the recombinase expression pattern. However well characterized it is, you should still test it yourself before using, just to be sure (Boxes 8.5 and 8.6).

Cre (or Flp) can also be introduced in vivo both in embryos and adults using a variety of DNA, RNA, or protein transfer methods, including virally mediated gene transduction, electroporation of tissues and organs, and exposure to cell-permeable recombinases. The advantages include the ability to choose the time and place to apply the recombinase; disadvantages include potential limitations of access to specific tissues and organs.

Box 8.5 Characterizing a *cre*-Transgenic or Knock-in Mouse Line

If you have generated transgenic or knock-in mouse lines to express Cre in specific tissues or have received Cre mice from another laboratory, you will need to characterize these animals to verify the tissue-specific pattern and timing of Cre activity. Here is a guide for the characterization of *cre*-transgenic or knock-in mouse lines (the same strategy can be used to characterize *flp* transgenic or knock-in mouse lines): If you have received the mice from another laboratory, the first thing you must do is genotype the mice to make sure that you have received the correct animals! The next step is to examine Cre expression. This can be performed at the level of the RNA (i.e., RT-PCR, northern, or in situ hybridization) or protein (immunohistochemistry).

Alternatively, you can use a Cre reporter mouse line such as *R26R* (Figure 8.5 and see Appendix 1). Typically, Cre-expressing mouse lines are crossed with Cre reporter mouse lines to generate double heterozygotes, and the reporter is used to visually assess the temporal and spatial patterns of Cre activity. Examine reporter expression in double heterozygotes at various developmental stages, either in whole mounts or histological sections. It is very important to determine the first stages of development in which you detect Cre activity because this indicates when and in

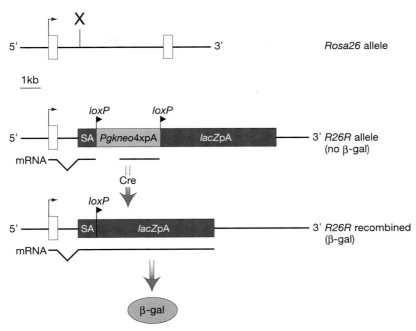

Figure 8.5. Generation and use of the *R26R* Cre reporter mouse line. The *Rosa26* locus encodes transcripts that lack an open reading frame. The *R26R* allele was generated by targeting a cassette into the *XbaI* (X) site located within the first intron. A *neo* expression cassette with 4xpA sequences blocks the transcription of the *lacZ* gene. Cre recombinase will delete the floxed *neo* cassette, allowing *lacZ* transcription. The splice acceptor (SA) sequence leads to the generation of a chimeric *R26-lacZ* transcript that results in the production of β-galactosidase (β-gal). All of the cellular progeny derived from a cell that has undergone this Cre-mediated recombination will stain positively for β-gal activity.

which tissues a conditional knockout might first show its effects. Keep in mind that the expression of the reporter is the result of Cre-mediated recombination in a given cell at a specific time and that all of the cellular progeny of that cell will continue to express the reporter (fate mapping). Thus, not all reporter-positive cells necessarily express the recombinase. It should be noted that some Cre reporters might be more sensitive to Cre activity. Therefore, you may want to characterize your Cre mouse lines with more than one type of Cre reporter mouse.

Mosaic expression of a *cre* transgene or knock-in is relatively common. Accessing the amount and variation of mosaicism will be important for anticipating potential mutant phenotypes in a conditional knockout experiment. Genetic background can influence, either positively or negatively, the expression of a *cre* transgene or knock-in. Therefore, it is useful to characterize the expression of your *cre* transgene or knock-in on different genetic backgrounds or at least on the background you intend to use.

It is possible that a tissue-specific *cre* transgene or knock-in will also express Cre in the male or female germ line, i.e., in the developing or mature germ cells. To determine if this is the case, generate mice that are heterozygous for the *cre* transgene or knock-in and also heterozygous for a Cre reporter allele. Then breed these double heterozygous mice to wild-type mice and examine Cre reporter expression in the resulting progeny. If Cre is active in the germ line, ubiquitous expression of the Cre reporter will be present in the progeny. If you do find germ-line Cre activity in a tissue-specific Cre line, it may be limited to only the male or female germ line, in which case you can adjust your crosses accordingly using the germ line that does not have Cre activity.

Finally, remember that the timing of Cre activity and activation of a Cre reporter will likely be different from the timing of loss of expression of the endogenous gene by a conditional knockout (Figure 8.6).

Figure 8.6. Comparison of Cre reporter activity with Cre-mediated loss of function at the cellular level. (*A*) Detection of β-gal activity in *R26R/+* × *cre* mice. As a *cre* transgene is transcribed and translated, Cre activity increases. At a certain threshold of Cre activity, the floxed DNA that blocks β-gal expression is deleted (*dashed arrow*). *lacZ* is then transcribed and translated into β-gal. At a certain time, there will be sufficient β-gal activity and individual cells will stain blue with X-gal (*shaded area*). (*B*) Development of a deficiency of protein activity caused by Cre-mediated recombination of a conditional null allele (flox). As in *A* above, as the *cre* transgene is transcribed and translated, Cre activity increases. At a certain threshold of Cre activity, the floxed DNA is deleted (*dashed arrow*). Because the cell genotype is flox/–, no further gene product can be generated from the target gene. Residual target gene mRNA and protein will decay, continuously reducing residual levels of target gene protein activity. Once protein activity levels drop below a threshold (*dashed line*), the cell will be functionally deficient for that protein (*shaded area*). Thus, while a Cre reporter provides information about when Cre activity first initiates, it does not necessarily tell you when your target gene product will be sufficiently reduced to cause a functional consequence.

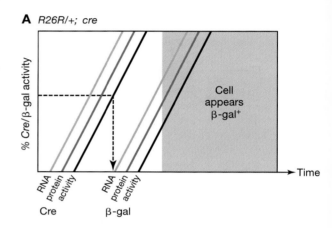

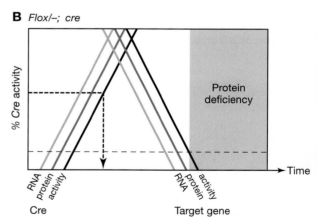

With respect to verifying which cells have excised the floxed/flrted exons, unless you have built a reporter into the allele that reports when the exons are excised (see Figure 2.5B), you will have to rely on assumptions based on the known pattern of recombinase expression or on direct measurements of either Cre/Flp expression or the absence of the floxed/flrted exons. Because the level of recombinase expression may vary for different transgenes and the time of onset of recombinase expression may be gradual (Figure 8.6), you will have to consider the

> **Box 8.6** Characterizing Inducible Cre Estrogen Receptor or Cre Progesterone Receptor Transgenic or Knock-in Lines
>
> Cre estrogen receptor (ER) or Cre progesterone receptor (PR) and their modified versions are fusion proteins composed of Cre and mutant ligand binding domains of the ER and PR, respectively. These fusion proteins do not bind estrogen or progesterone but rather tamoxifen and RU-486, respectively. In the absence of these artificial ligands, the Cre fusion proteins reside in the cytoplasm. However, in their presence, the Cre fusion proteins enter the nucleus and are active.
>
> The characterization of a CreER or CrePR transgenic or knock-in mouse line is similar to that described for a constitutively active Cre mouse line (Box 8.5). However, in addition to crosses with Cre reporter mice, you also need to test the inducibility and extent of Cre activity at different times after administration of the inducing drug. This can require a lot of time, so if you have generated new transgenic mouse lines with tissue-specific inducible Cre expression, it may be more practical to first assay for Cre expression at the level of RNA or protein. Once you have identified expressing transgenic lines, then generate double heterozygotes with the inducible *cre* transgene and a Cre reporter mouse to test for inducibility.
>
> The analysis of embryonic stages requires treatment of a pregnant female carrying *cre* transgene and Cre reporter double-heterozygous embryos with tamoxifen (for CreER) or RU-486 or its analogs (for CrePR) at different stages of gestation. You will need to determine the sufficient drug dose and the number and timing of injections needed during gestation for the excision of the floxed allele and loss of gene product. These drugs are light sensitive, so take precautions. In addition, it is very important to have the drugs completely dissolved in the diluent, which may require overnight stirring. For drug injections into mice, use vegetable oil as the diluent. Daily injections of the drug may be necessary to achieve your desired results. You can also send out tamoxifen or RU-486 for encapsulation. Though expensive, these capsules can then be implanted under the skin of the animal for constant slow release of the drug. Some inducible Cre lines express in the skin. These drugs can be topically applied, dissolved in 100% ethanol.
>
> Below is a guide for suggested starting doses of tamoxifen and RU-486 for induction of Cre activity. You will have to optimize these for your particular study.
>
> - Tamoxifen dose for injection of pregnant females is 0.5–10 mg/pregnant female.
> - Tamoxifen dose for injection of postnatal animals is 0.75–1 mg for 5 consecutive days.
> - RU-486 (24-hour half life) dose for injection of pregnant females is 50–100 µg/kg body weight; to prevent abortion, RU-486-treated mice must be injected daily with progesterone at 0.5 mg/mouse.
> - RU-486 dose for injection of postnatal animals is 100–150 µg/kg body weight, probably multiple (three) daily injections.
>
> These doses can serve as a starting point for your particular study. Examine the fetuses for the expression of the reporter at different times after the treatment. The analysis of postnatal stages requires treatment of the double heterozygotes after birth. Again, you will need to determine the drug dose and the number and timing of injections to elicit sufficient recombination for your studies. After the induction of Cre activity, analyze the tissue of interest for expression of the Cre reporter at different times after the treatment to assess the timing and extent of Cre activity. These data can then be used as a guide for your intended inducible conditional knockout.

possibility that tissues are mosaic for excision of the exons in the offspring of the cross and interpret the results accordingly.

Once you have chosen a suitable recombinase transgenic line, tissue- or stage-specific mutations can be generated using several breeding schemes. A very efficient scheme is to generate males heterozygous for a constitutive null allele of your gene and heterozygous or hemizygous for the targeted knock in or transgene that directs tissue-specific recombinase expression, Cre in this example, +/–; cre/+ (Figure 8.7). The constitutive null allele can either be an independent null allele or a recombined null allele generated from the conditional null allele. Only a few +/–; cre/+ males are needed for crosses with females homozygous for the conditional null allele (*flox/flox*). It is likely that you will need to sacrifice the *flox/flox* females to analyze embryonic time points, but they are easy to generate from a homozygous breeding stock, which obviates the need for genotyping. Using this cross, one in four of the resulting progeny will be *flox/–; cre/+*. With this strategy, Cre only has to act on one allele because the other is already null. If Cre is required to act on two floxed alleles per cell, it might not be as efficient as expected and may create mosaicism for the tissue-specific knockout. This in turn may generate a less severe mutant phenotype than a null.

One note of caution: If you are using a recombinase "knock-in" allele for tissue-specific ablation (i.e., a recombinase gene that is inserted into an endogenous locus under the control of an endogenous promoter), it is also likely to be a loss-of-function allele for the gene into which the recombinase is inserted and could potentially influence the expressivity and penetrance of your tissue-specific mutant phenotype. The only ways around this are fairly drastic (e.g., recreating the recombinase-expressing mouse without ablating the endogenous gene

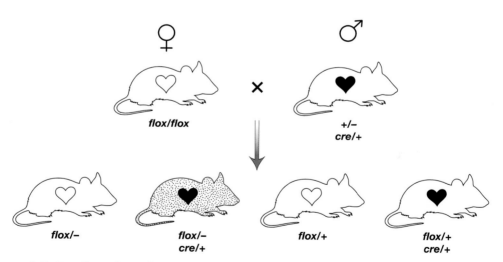

Figure 8.7. Breeding scheme for producing mice or embryos with a tissue-specific mutation. Females homozygous for a conditional allele (*flox*) are mated with males heterozygous or hemizygous for a gene-specific *cre* knock in or a tissue-specific *cre* transgene (*cre/+*)—in this case, a heart-specific *cre* (*solid heart*) and heterozygous for a null allele of the gene (+/–). One in four of the resulting offspring (*stippled*) will be heterozygous for the two mutant alleles (*flox/–*) and also contain the *cre* transgene (*cre/+*), in which case the floxed exons will be deleted from the conditional allele in the tissues expressing Cre, namely, the heart, resulting in a heart-specific null.

function or using a comparable recombinase transgenic strain if one is available), but you should keep this in mind when interpreting your results. Similarly, you may have little choice of genetic background when it comes to selecting a recombinase-expressing mouse to breed with your conditional allele. Short of making congenic strains (see Box 4.4, p. 54), there is little you can do but accept that you are dealing with a mixed genetic background in the analysis (Section 4.B).

Ideally, you should be able to document the loss of protein from your targeted allele in the recombinase-expressing tissue. In many cases, however, antibodies may not be available. Documenting the loss of mRNA for the gene is the next best strategy and could be done on a cell-by-cell basis with in situ hybridization using an exon-specific probe within the floxed exons (see Section 2.B.4) or on a whole-tissue basis using RT-PCR and exon-specific primers. It is important to realize that the timing of recombinase expression and the loss of mRNA, protein, and eventually function will be gradual. Timing will depend on the rate of turnover of the protein in question and, for inducible recombinase expression, the timing and kinetics of induction (Figure 8.6).

Finally, the analysis. Knowing when and where the specific recombinase is expressed, you can extrapolate to a projected time of effect of gene ablation on a given tissue. Now, adapt the methods outlined in Chapter 5 and 6 to study the phenotype produced by the tissue or stage-specific ablation of gene function.

HELPFUL HINT

When you generate +/–; cre/+ stud males for crosses with flox/flox females in a tissue-specific knockout experiment (see Figure 8.7), prescreen the males by crossing them with Cre reporter females (e.g., R26R) to verify that they can produce progeny with the correct pattern of Cre activity. This is especially important when you know that the tissue-specific pattern of Cre activity for the particular cre transgenic line that you are using is sensitive to genetic background, position effects, or stochastic variation.

8.D Regulatory Mutations

The mouse genome has been sequenced, assembled, and annotated and we now know the identity of most, if not all, coding regions. By comparison, relatively little is known about the specific sequences that regulate gene transcription and other gene regulatory processes. Empirical studies of sequences located near coding regions have traditionally been used to identify tissue-specific transcriptional enhancers. Because of the availability of multiple assembled animal genomes, cross-species comparisons of noncoding regions have begun to identify conserved sequences as candidate transcriptional regulatory elements that are then being tested using transgenic assays. Thus, more and more tissue-specific transcriptional enhancers are being identified. This opens up the possibility of creating tissue-specific loss-of-function alleles by targeted mutagenesis of known tissue-specific transcriptional enhancers (Figure 8.8). These regulatory mutations are not limited to transcriptional enhancers; they could also include mutations that affect RNA splicing, polyadenylation sequences, RNA/protein binding sites in untranslated regions, etc. Just like a Cre/*loxP*-based conditional knockout, a regulatory mutation can allow you to bypass an early lethality caused by a standard knockout.

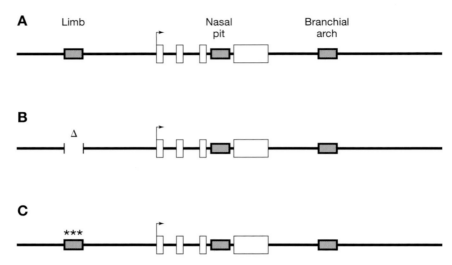

Figure 8.8. Tissue-specific knockouts caused by mutations in tissue-specfic enhancers. (*A*) Structure of a gene with tissue-specific enhancers (*shaded boxes*) for limb, nasal pit, and branchial arch expression distributed 5′, within, and 3′ of the exons (*open boxes*), respectively. (*B*) Targeted deletion (Δ) of the limb enhancer. (*C*) Targeted base-pair changes (***) in a transcription factor binding site of the limb enhancer that eliminates binding. The alleles shown in *B* and *C* should result in loss of transcription of the gene only in the limb resulting in a tissue-specific knockout, unless there is a redundant limb enhancer.

Regulatory mutations can generally be used for genes that are expressed in multiple tissues but not ubiquitously. This implies that the gene in question is regulated by multiple and essential transcriptional enhancers. If the regulatory sequence acts redundantly with another *cis* element, then it may contribute only partially to the overall levels of tissue-specific transcription. Therefore, mutation of a redundant element may result in a hypomorphic allele that can also be very useful (see Section 4.E.2). Regulatory mutations can be made using a variety of gene-targeting methods (i.e., Cre/*loxP*, double replacement, hit and run; see Section 2.B). Very simply, base-pair changes are engineered to disrupt the enhancer activity of a regulatory sequence. These base-pair changes prevent the binding of *trans*-acting factors that regulate tissue-specific transcription. As with any mutation, regulatory mutations can be made homozygous or combined with a null allele, if available.

A tissue-specific knockout caused by a regulatory mutation differs fundamentally from a traditional Cre/*loxP*-based tissue-specific knockout in a number of ways. First, only one mutant allele (i.e., one mouse strain) is required for a regulatory mutation approach in comparison to at least two but more likely three alleles required for a Cre/*loxP*-based tissue-specific knockout. In addition, with a Cre/*loxP*-based system, the loss of the target gene product depends upon the timing, action, and extent of Cre expression in the tissue of interest. With regulatory mutation, transcription will not be initiated in the tissue of interest, and, thus, an immediate deficiency of the gene product results.

APPENDIX 1

Useful Books and Resources

Absolute Essentials

Joyner A.L., ed. 2000. *Gene Targeting: A Practical Approach*, 2nd edition. Oxford University Press, United Kingdom.
   An indispensable practical tool and information resource for scientists generating mouse mutants using embryonic stem (ES) cell technology.

Kaufman M.H. 1992. *The Atlas of Mouse Development*. Academic Press, New York.
   The standard reference work and the definitive account of mouse embryology and development including serial histological sections of embryos throughout development, scanning electron micrographs of the external appearance of embryos, and special sections on the development of specific organ systems such as heart, tongue, palate, gonads, and kidney.

Nagy A., Gertsenstein M., Vintersten K., and Behringer R., eds. 2003. *Manipulating the Mouse Embryo: A Laboratory Manual*. Cold Spring Harbor Laboratory Press, Cold Spring Harbor, New York.
   The premier source of technical and theoretical guidance for mouse developmental biologists and geneticists; contains all the commonly used protocols for embryo and gene manipulation.

Opening artwork, *Gakusyu Nezumi*, painted by Kyosai Kawanabe (1831–1889), Freer Gallery of Art, Smithsonian Institution, Washington, D.C. Purchase F1975.29.9.

Rossant J. and Tam P.P.L., eds. 2002. *Mouse Development: Patterning, Morphogenesis, and Organogenesis*. Academic Press, San Diego, California.
> A valuable collection of highly readable, informative reviews of mouse development from fertilization through axis formation and gastrulation, to lineage specification and differentiation, to detailed organogenesis of many of the major organ systems. Includes up-to-date research results.

Silver L.M. 1995. *Mouse Genetics: Concepts and Applications*. Oxford University Press, United Kingdom.
> A comprehensive practical guide to mouse breeding and genetics and an introduction to the mouse as a model system for genetic analysis; contains a complete description of the laboratory mouse, the tools used in analysis, procedures for carrying out genetic studies, and a detailed guide for performing breeding studies and interpreting experimental results. Full text is available online at http://www.informatics.jax.org/silver/

Mouse Genetics and Husbandry

Green E.L., ed. 1998. *Biology of the Laboratory Mouse, by the Staff of the Jackson Laboratory*, 2nd revised edition. Dover Publications, New York.
> A golden oldie that is a comprehensive treatment of all aspects of mouse biology. This title emphasizes genetic variations of mice and their exploitation for the solution of biological problems. Includes practical chapters on husbandry, breeding systems, and record keeping, as well as chapters on development, physiology, life span, susceptibility to disease and tumors, and much more.

Jackson Laboratory. 1991. *Handbook on Genetically Standardized JAX Mice*, 4th edition. JAX Mice Literature, The Jackson Laboratory, Bar Harbor, Maine.
> A highly detailed compendium of information on all aspects of growth, reproduction, life span, and disease susceptibility of the various inbred, hybrid, recombinant inbred, congenic, mutant strains and stocks, and mice with chromosomal aberrations available from JAX. Also contains general husbandry procedures.

Lyon M.F., Rastan S., and Brown S.D.M., eds. 1995. *Genetic Variants and Strains of the Laboratory Mouse*, 3rd edition. International Committee on Standardized Genetic Nomenclature for Mice, Oxford University Press, New York.
> A two-volume reference on the fundamentals of mouse genetics for researchers and advanced students; catalogs and describes known genetic variants and inbred strains of the mouse.

Silvers W.K. 1979. *The Coat Colors of Mice: A Model for Mammalian Gene Action and Interaction*. Springer-Verlag, Heidelberg, Germany.
> Describes the many genetic factors that influence the coat colors of the house mouse and illustrates various multidisciplinary approaches that have been used to explore genotype-

phenotype pathway; contains information on a large number of specific alleles at different loci and how they affect pigmentation.

Suckow M.A., Danneman P., and Brayton C. 2001. *The Laboratory Mouse*. A volume in the *Laboratory Animal Pocket Reference Series*. CRC Press, Boca Raton, Florida.
   Contains the basics of management, husbandry, equipment, and regulations involved in maintaining an animal facility, plus useful information on diseases and veterinary care of mice.

Development, Developmental Landmarks, and Staging Systems

Downs K.M. and Davies T. 1993. Staging of gastrulating mouse embryos by morphological landmarks in the dissecting microscope. *Development* **118:** 1255–1266.

Fujinaga M., Brown N.A., and Baden J.M. 1992. Comparison of staging systems for the gastrulation and early neurulation period in rodents: A proposed new system. *Teratology* **46:** 183–190.

Kaufman M.H. and Bard J.B.L. 1999. *The Anatomical Basis of Mouse Development*. Academic Press, New York.
   A companion to Kaufman's *The Atlas of Mouse Development*, which details the developmental anatomy of the early embryo, the transitional tissues, and all the major organ systems. It includes extensive reference indexes detailing developmental stage criteria, when tissues first appear, and the constituent tissues of embryos at each stage of development. Also covered are comparisons with normal and abnormal human development. Useful for understanding developmental anatomy in normal and mutant mice.

Rugh R. 1990. *The Mouse: Its Reproduction and Development*. Oxford University Press, United Kingdom.
   A well-illustrated text on mouse development including reproduction, chronology of normal development, and organogenesis; includes histological sections and gross anatomy with many useful morphological markers of different stages of development.

Theiler K. 1989. *The House Mouse: Atlas of Embryonic Development*, 2nd edition. Springer-Verlag, New York.
   In this classic compendium, the author presents each stage of mouse development with photographs and micrographs. Organ systems are systematically reconstructed from fertilization until after birth.

Wanek N., Nuneoka K., Holler-Dinsmore G., Burton R., and Bryant S.V. 1989. A staging system for mouse limb development. *J. Exp. Zool.* **249:** 41–49.

Methods

Copp A.J. and Cockroft D.L., eds. 1990. *Postimplantation Mammalian Embryos: A Practical Approach*. Oxford University Press, United Kingdom.

Methods of observing, manipulating, and analyzing implanted embryos. The topics include exo utero surgery, the morphological stages of postimplantation embryonic development, culturing postimplantation embryos, extracting macromolecules, and teratogen testing.

Jackson I.I. and Abbott C.M., eds. 2001. *Mouse Genetics and Transgenics: A Practical Approach*. Oxford University Press, United Kingdom.
Covers all aspects of using the mouse as a genetic model organism: care and husbandry; archiving stocks as frozen embryos or sperm; making new mutations by chemical mutagenesis; transgenesis; gene targeting; mapping mutations and polygenic traits by cytogenetic, genetic, and physical means; and disseminating and researching information via the Internet.

McLaren A., ed. 1976. *Mammalian Chimaeras*. Developmental and cell biology series, Cambridge University Press, United Kingdom.
A classic monograph on what chimeras are, how to make them, and how to use them.

Pinkert C.A., ed. 2002. *Transgenic Animal Technology: A Laboratory Handbook*. Academic Press, Boston, Massachusetts.
Covers the technical aspects of gene transfer—from molecular methods to whole animal considerations—for important laboratory and domestic animal species.

Robertson E.J., ed. 1987. *Teratocarcinomas and Embryonic Stem Cells: A Practical Approach*. Oxford University Press, United Kingdom.
Contains the definitive work on derivation and culture of ES cells. A classic.

Stern C.D. and Holland P.W.H., eds. 1993. *Essential Developmental Biology: A Practical Approach*. Oxford University Press, United Kingdom.
Comprehensive, easy-to-follow protocols, and practical instructions for techniques, from traditional embryology to cellular and molecular methods. Includes complete reprints of stage tables in common use for the main laboratory species.

Tuan R.S. and Lo C.W., eds. 2000. *Developmental Biology Protocols*, volumes I–III. *Methods Mol. Biol.*, vol. 135–137. Humana Press, Totowa, New Jersey.
Three-volume laboratory manual providing principles, background, rationale, and step-by-step instructions for studying and analyzing the events of embryonic development. Covers developmental pattern and morphogenesis, embryo structure and function, cell lineage analysis, chimeras, experimental manipulation of embryos, application of viral vectors, organogenesis, abnormal development and teratology, screening and mapping of novel genes and mutations, transgenesis production and gene knockout, manipulation of developmental gene expression and function, analysis of gene expression, models of morphogenesis and development, and in vitro models and analysis of differentiation and development.

Tymms M.J. and Kola I., eds. 2001. *Gene Knockout Protocols*. Humana Press, Totowa, New Jersey.
Describes techniques for the design of gene-targeting constructs and analysis of the mouse phenotype, covering techniques of embryo transplantation, in vitro embryonic stem cell

differentiation, creation of aggregation chimeras, mouse pathology, embryo cryopreservation, and transplantation.

Wassarman P.M. and DePamphilis M.L., eds. 1993. *Guide to Techniques in Mouse Development. Methods Enzymol.*, vol. 225, Academic Press, New York.
An extremely valuable, comprehensive compendium of techniques used in the study of mouse development, including record keeping for mouse colonies, in vitro fertilization, preimplantation and postimplantation embryo culture, and analysis of gene expression including in situ hybridization, gene targeting, and making transgenics.

Wilkinson D.G., ed. 1998. *In Situ Hybridization: A Practical Approach*, 2nd edition. Oxford University Press, United Kingdom.
Protocols detailing the major techniques of in situ hybridization including in situ hybridization to mRNA with oligonucleotide and RNA probes; analysis using light and electron microscopes; whole-mount in situ hybridization; double detection of RNAs and RNA plus protein; and fluorescent in situ hybridization to detect chromosomal sequences.

Phenotypic Analysis

Conway S.J., Kruzynska-Frejtag A., Kneer P.L., Machnicki M., and Koushik S.V. 2003. What cardiovascular defect does my prenatal mouse mutant have, and why? *Genesis* **35:** 1–21.
Review of phenotypic analysis of embryos with heart problems.

Crawley J.N. 2000. *What's Wrong With My Mouse? Behavioral Phenotyping of Transgenic and Knockout Mice.* Wiley-Liss, New York.
A molecular geneticists' guide to the best tests for analyzing behavorial phenotypes in transgenic and knockout mice.

Maronpot R., Boorman G.A., and Gaul B.W., eds. 1999. *Pathology of the Mouse: Reference and Atlas.* Cache River Press, Vienna, Illinois.
Provides a comprehensive survey of mouse pathology; a reference and atlas in that it covers all systems of the body and is heavily illustrated.

Rogers D.C., Fisher E.M.C., Brown S.D.M., Peters J., Hunter A.J., and Martin J.E. 1997. Behavioral and functional analysis of mouse phenotype: SHIRPA, a proposed protocol for comprehensive phenotype assessment. *Mamm. Genome* **8:** 711–713.
A procedure to characterize the phenotype of mice in three stages: a primary screen utilizing standard methods to provide a behavioral and functional profile by observational assessment, a secondary screen involving a comprehensive behavioral assessment battery and pathological analysis, and a tertiary screen tailored to the assessment of existing or potential models of neurological disease as well as assessment of phenotypic variability.

Sundberg J.P. 1994. *Handbook of Mouse Mutations with Skin and Hair Abnormalities: Animal Models and Biomedical Tools.* CRC Press, Boca Raton, Florida.
Overview of mutant mouse resources and use, major features of cutaneous biology, and comparison of phenotypes of mouse mutations with human diseases.

Sundberg J.P. and Boggess D., eds. 2000. *Systematic Approach to Evaluation of Mouse Mutations*. CRC Press, Boca Raton, Florida.

> Geared toward evaluating spontaneous mutations in adult mice using scaled-down versions of standard operating procedures for diagnostic medicine and biomedical research. Includes chapters on necropsy, photography, colony establishment and management, and JAX mouse resources for mutant mice.

Ward J.M., Mahler J.F., Maronpot R.R., Sundberg J.P., and Frederickson R.M., eds. 2000. *Pathology of Genetically Engineered Mice*. Iowa State University Press, Ames.

> Approaches to organ-specific evaluations, with examples of many currently available genetically engineered and naturally mutant mice to illustrate how specific defects can be studied. Topics include using confocal microscopy to study transgenic and knockout mice, inducing mutations, genetic background effects on the interpretation of phenotypes in induced mutant mice, the pathologic characterization of neurological mutants, and hepatic pathology.

Web-based Resources

Archiving mutants

- NIH Mutant Mouse Regional Resource Centers (MMRRC)
  http://www.nih.gov/science/models/mouse/resources/mmrrc.html

- Induced Mutant Resource (IMR)
  http://www.jax.org/imr/index.html

- European Mouse Mutant Archive (EMMA)
  http://www.emma.rm.cnr.it

Conditional genetic resource

Compendium of Cre, floxed, and Cre-excision reporter lines.
http://www.mshri.on.ca/nagy/

Frimorfo

Company specializing in the histological characterization of genetically modified rodents; includes a collection of images of normal mouse tissue, the Morphome Database.
http://www.frimorfo.com

IGTC international gene trap consortium

Generating an international resource of embryonic stem cells with gene trap insertions in all genes in the mouse genome.
http://www.igtc.ca/

Mouse Genome Informatics (MGI)

www.informatics.jax.org

Provides integrated access to data on the genetics, genomics, and biology of the laboratory mouse including the following:

- Mouse Genome Database (MGD) includes data on gene characterization, nomenclature, mapping, gene homologies among mammals, sequence links, phenotypes, allelic variants and mutants, and strain.

- Gene Expression Database (GXD) integrates different types of gene expression information from the mouse and provides a searchable index of published experiments on endogenous gene expression during development.

- Mouse Genome Sequence (MGS). The goal of the MGS project is to integrate emerging mouse genomic sequence data with the genetic and biological data available in MGD and GXD. MGS supports the MouseBLAST server as a sequence-level entry point into the MGI Database.

- Mouse Tumor Biology (MTB) Database integrates data on the frequency, incidence, genetics, and pathology of neoplastic disorders, emphasizing data on tumors that develop characteristically in different genetically defined strains of mice.

- Gene Ontology (GO). The Mouse Genome Informatics (MGI) group is a founding member of the Gene Ontology Consortium (www.geneontology.org). MGI fully incorporates GO in the database and provides a GO browser.

Mouse Nomenclature

- The authoritative source of official names for mouse genes, alleles, and strains. Nomenclature follows the rules and guidelines established by the International Committee on Standardized Genetic Nomenclature for Mice and is implemented through the Mouse Genomic Nomenclature Committee (MGNC).
    http://www.informatics.jax.org/mgihome/nomen/index.shtml

- New nomenclature for strain 129 mice.
    http://www.informatics.jax.org/mgihome/nomen/strain_129.shtml

Mouse 3D Atlas

http://genex.hgu.mrc.ac.uk

NCI Mouse Models of Human Cancers Consortium (MMHCC)

A collaborative program designed to derive and characterize mouse models and to generate resources, information, and innovative approaches to the application of mouse models in cancer research.
http://emice.nci.nih.gov/emice/mouse_models

Transgenic/Targeted Mutation Database (TBASE)

Information on transgenic mice and targeted mutations including phenotypes.
http://tbase.jax.org/

APPENDIX 2

Examples of Phenotype Analysis

EXAMPLES AND KEY REFERENCES for specific types of phenotypic analyses or analyses using specific techniques are keyed to specific sections of the book. The references are not meant to be comprehensive, but rather provide a cogent example of different situations and how we or other investigators have dealt with them.

Section	Situation or phenotype	Example	Reference
1.A	Two classes of heterozygous females depending on maternal or paternal inheritance of mutant allele due to nonrandom X inactivation in placenta	*Esx1*	Li and Behringer (1998)
1.B.1	ENU mutagenesis screen	dominant screen	Hrabe de Angelis et al. (2000)
		recessive screen	Kasarskis et al. (1998)
		balancer screen	Kile et al. (2003)
1.B.2	X-ray mutagenesis screen	albino deletions	Holdener and Magnuson (1994)
1.B.3	Insertional mutagenesis	gene trap	Gossler et al. (1989)
		viral	Friedrich and Soriano (1991)
		transposon	Horie et al. (2003)
1.B.4	Transgenesis	*Gh*	Palmiter et al. (1982)
2.A.2	Selectable marker cassette affecting gene expression at targeted locus	*Myf5*	Kaul et al. (2000)
	Marker recycling to reuse a targeting vector for the second allele	*Msh3*	Abuin and Bradley (1996)
Box 2.2	HSV *tk* and male infertility		Braun et al. (1990)
2.B.1	Regulatory element for a gene in an intron of a neighboring gene	*Shh*	Lettice et al. (2003)
2.B.2	An expression reporter producing a hypermorphic allele	*Vegfa*	Miquerol et al. (2000)
2.B.4	Conditional allele that acts as a null or hypomorphic allele	*Fgf8*	Meyers et al. (1998)
3.B.1.c	Mutant ES cells causing the death of chimeric embryos (X-linked gene)	*Gata1*	Pevny et al. (1991)
3.B.2.d	A coat color phenotype that results in good chimeras not showing strong coat color chimerism	*Pdgfra*	Soriano (1997)
3.C.2.b	HSV *tk* and male infertility		Braun et al. (1990)
3.C.2.d	No germ-line transmission from male chimeras due to a dominant effect of the mutation on gametogenesis	*Klhl10*	Yan et al. (2004)

Section	Situation or phenotype	Example	Reference
3.C.3.b	Male germ-line chimeras produce no heterozygous offspring due to dominant effect on embryos	Vegfa	Carmeliet et al. (1996)
3.C.3.c	Perinatal lethality due to a dominant effect in the heterozygous offspring of chimeras	Sox9	Bi et al. (2001)
3.D.b	Dominant effect on gametogenesis resulting in infertility	Klhl10	Yan et al. (2004)
3.D.c	Sex-limited dominant effect on fertility	Odsex	Bishop et al. (2000)
3.D.d	Dominant sex reversal	Odsex	Bishop et al. (2000)
Box 4.2	Different phenotype on different genetic backgrounds	Egfr	Threadgill et al. (1995)
4.D.1	Deletion of a selection cassette in vivo	Sox2Cre	Hayashi et al. (2003)
		CmvCre	Arango et al. (1999)
4.D.2	Maintaining a mutant with a balancer chromosome	Dph2l1	Chen and Behringer (2004)
4.D.3.a	Dealing with male infertility: intracytoplasmic sperm injection	Tnp1, Tnp2	Zhao et al. (2004)
4.D.3.b	Dealing with female infertility	Lif	Stewart et al. (1992)
4.E.1	Complementation testing to determine allelism of two mutations	Tbx6	Watabe-Rudolph et al. (2002)
4.E.2	Detection of a hypomorphic allele	Tbx6	Watabe-Rudolph et al. (2002)
4.E.3.a	Functional redundancy or compensation by genes in the same gene family	Myod1, Myf5	Rudnicki et al. (1993)
4.E.3.d	Similar mutant phenotypes but with no genetic interaction	Amh, Amhr2	Behringer et al. (1994), Mishina et al. (1996)
4.E.3.e	Mutant phenotype caused by over-expression of a downstream gene	Leftb, Cer1, Nodal	Perea-Gomez et al. (2002)
4.E.4	"Cheating Mendel"	Sox9	Akiyama et al. (2002)
5.C.1.a	No mutant preimplantation embryos due to a sperm/oocyte incompatibility	DDK syndrome	Babinet et al. (1990)
5.C.1.b	Arrest or delay during cleavage	Pes1	Lerch-Gaggl et al. (2002)
Box 5.5	Cell counting in ICM and trophectoderm of preimplantation embryos	Fgf4	Goldin and Papaioannou (2003)
5.C.1.c	Abnormal compaction of preimplantation embryos	Cdh1	Ohsugi et al. (1997)
5.C.1.d	Abnormal blastocyst morphology (other than compaction problem)	Nup214	van Deursen et al. (1996)
5.C.2.a	Failure to hatch	Smarc1b	Guidi et al. (2001)
5.C.2.b	Failure to attach	Mcl1	Rinkenberger et al. (2000)
5.C.2.c	Failure of blastocyst outgrowth	Trrap	Herceg et al. (2001)
	Failure of ICM in blastocyst outgrowth	Fgf4	Feldman et al. (1995)
5.C.3	Analysis of isolated inner cell mass	Mcl1	Rinkenberger et al. (2000)
5.C.4	Analysis of embryos in implantation delay	Lif	Nichols et al. (2001)
5.C.5.a	TUNEL on blastocysts	Mdm2	Chavez-Reyes et al. (2003)
	In situ on preimplantation embryos	Mcl1	Rinkenberger et al. (2000)
	Two-dimensional gels on preimplantation embryos		Shi et al. (1994)
5.C.5.b	Gene expression in whole pre-implantation embryos: in situ	Eomes	Hancock et al. (1999)
	Whole-mount immunohistochemistry on preimplantation embryos	Pou5f1	Liu et al. (2004)
	RT-PCR on preimplantation embryos	Fgf4	Goldin and Papaioannou (2003)
5.D.1	Periimplantation death 4.5–5.5 dpc	Mdm2	Montes de Oca Luna et al. (1995)
		Fgf4	Feldman et al. (1995)
5.D.2	Lethality 6.5–9.5 dpc: failure of gastrulation	Wnt3	Liu et al. (1999)

Section	Situation or phenotype	Example	Reference
	Lethality 6.5–9.5 dpc: failure of extra-embryonic membrane/placenta to function properly	*Dlx3*	Morasso et al. (1999)
	Lethality 6.5–9.5 dpc: failure of allantoic fusion	*Tbx4*	Naiche and Papaioannou (2003)
	Lethality 6.5–9.5 dpc: early growth retardation with normal developmental stage	*Myc*	Davis et al. (1993)
	Lethality 6.5–9.5 dpc: failure of vasculogenesis or hematopoiesis	*Jag1*	Xue et al. (1999)
	Lethality 6.5–9.5 dpc: cardiac problems	*Mef2c*	Lin et al. (1997)
	Lethality 6.5–9.5 dpc: heart looping defects	*Mef2c*	Lin et al. (1997)
	Neural tube closure defects	*Twist1*	Chen and Behringer (1995)
		Cart1	Zhao et al. (1996)
		Shrom	Hildebrand and Soriano (1999)
	Laterality defects	*Invs*	Morgan et al. (1998)
	Genetic interactions in early-gestation lethals	*Brca1, Brca2, Trp53*	Ludwig et al. (1997)
		Mdm2, Trp53	Montes de Oca Luna et al. (1995)
5.D.2.a	Genotyping embryos from histological sections	*Hdh*	Zeitlin et al. (1995)
5.D.2.b	Molecular characterization of mutant embryos	*Tbx6*	Chapman and Papaioannou (1998)
5.D.2.c	Alterations in proliferation and/or apoptosis	*Hdh*	Zeitlin et al. (1995)
5.D.2.d	In vitro culture of organs from early-gestation embryos:		
	eye	*Bmp4*	Furuta and Hogan (1998)
	allantois		Downs et al. (2000)
	Use of teratomas to investigate a mutant phenotype	*Tbx6*	Chapman et al. (2003)
	Using ES cells to investigate a mutant phenotype	*Twist*	Chen and Behringer (1995)
	Testing developmental potential of mutant cells in chimeras	*Tbx6*	Chapman et al. (2003)
5.E.1.a	Lethality 9.5 dpc–term: vasculogenesis and hematopoietic defects	*Smad5*	Chang et al. (1999)
5.E.1.b	Lethality 9.5 dpc–term: cardiovascular defects	*Tbx1*	Jerome and Papaioannou (2001)
		Tbx2	Harrelson et al. (2004)
	Lethality 9.5 dpc–term: yolk sac defect	*Tbx3*	Davenport et al. (2003)
	Lethality 9.5 dpc–term: placental defects	*Vhlh*	Gnarra et al. (1997)
	Separation of yolk sac germ layers		Yoder et al. (1994)
		Nodal	Varlet et al. (1997)
5.E.1.c	Lethality 9.5 dpc–term: placental insufficiency	*Dlx3*	Morasso et al. (1999)
	Large placenta, small fetus	*Esx1*	Li and Behringer (1998)
5.E.2	9.5 dpc–term: specific morphological defects:		
	skeleton	*Tbx1*	Jerome and Papaioannou (2001)
		Tbx3	Davenport et al. (2003)
	kidney	*Lhx1*	Shawlot and Behringer (1995)
	lung and limbs	*Fgf10*	Sekine et al. (1999)
	heart	*Tbx2*	Harrelson et al. (2004)
5.E.2.c	9.5 dpc–term: cell death	*Fos*	Roffler-Tarlov et al. (1996)
	Lethality 9.5 dpc–term: cell proliferation	*Foxc2*	Winnier et al. (1997)

Section	Situation or phenotype	Example	Reference
5.E.2.d	9.5 dpc–term: in vitro analysis of cells and organs:		
	derivation and use of mouse embryonic fibroblasts	Trp53, Mdm2	de Rozieres et al. (2000)
	mammary glands		Heuberger et al. (1982)
	lung buds		Miura and Shiota (2000)
	kidney primordia		Srinivas et al. (1999)
	limb buds	Tbx4	Naiche and Papaioannou (2003)
	heart		Kruithof et al. (2003)
	tooth buds	Msx1	Bei et al. (2000)
	gonad	Sox9	Chaboissier et al. (2004)
	palate	Tgfb3	Dudas et al. (2004)
	9.5 dpc–term: transplantation of cells and organs:		
	mammary buds	Igfr1	Bonnette and Hadsell (2001)
	spermatogonial stem cells		Brinster and Zimmermann (1994)
	skin	Pdgfa	Karlsson et al. (1999)
	testis		Schlatt et al. (2003)
	liver		Shiojiri et al. (1995)
	ovaries	Gsc	Rivera-Perez et al. (1995)
6.A.1.a	Perinatal lethality: catastrophic	Lhx1	Shawlot and Behringer (1995)
6.A.1.b	Perinatal lethality: cardiovascular	Tbx1	Jerome and Papaioannou (2001)
6.A.1.c	Perinatal lethality: developmental delay	Dph2ll	Chen and Behringer (2004)
6.A.1.d	Perinatal lethality: cranial nerve defects	Hoxa1	Lufkin et al. (1991)
6.A.1.e	Perinatal lethality: skeletal defects	Sp7	Nakashima et al. (2002)
6.A.1.f	Perinatal lethality: cleft palate	Msx1	Satokata and Maas (1994)
6.A.1.g	Perinatal lethality: diaphragm	Myog	Hasty et al. (1993)
6.A.1.h	Perinatal lethality: breathing	Bdnf	Erickson et al. (1996)
6.B.1.a	Postnatal phenotypes: before weaning, altered growth	Fos	Johnson et al. (1992)
	Postnatal phenotypes: before weaning, morphological	Egfr	Fitch et al. (1993)
	Postnatal phenotypes: before weaning, coordination/movement	Mpz	Giese et al. (1992)
6.B.1.b	Postnatal phenotypes after weaning: growth and vigor	Src	Soriano et al. (1991)
	Postnatal phenotypes after weaning: circadian rhythm alterations	Fos	Honrado et al. (1996)
	Postnatal phenotypes after weaning: neurological/behaviorial problems	Fos	Paylor et al. (1994)
	Postnatal phenotypes after weaning, infertility: female	Lif	Stewart et al. (1992)
	Postnatal phenotypes after weaning, infertility: male	Sys	MacGregor et al. (1990)
	Postnatal phenotypes after weaning: reproductive behavior in males	Fos	Baum et al. (1994)
6.B.1.c	Age-dependent phenotypes: tumors	Trp53	Donehower et al. (1992)
	Age-dependent phenotypes: rheumatoid arthritis	Il1r1	Horai et al. (2000)
6.B.2.a	No visible mutant phenotype	Hprt	Kuehn et al. (1987)
	Test of vision	Vsx1	Ohtoshi et al. (2004)
	Test of hearing	Ysb	Dong et al. (2002)
	Test of touch	Drg11	Chen et al. (2001)
6.B.2.b	Neurological tests of balance and coordination	C57BL6 and 129 strains	Kelly et al. (2003)

Section	Situation or phenotype	Example	Reference
6.B.2.c	Genetic challenge of overtly normal mutants	*MyoD, Myf5*	Rudnicki et al. (1993)
6.B.2.d	Environmental challenge of overtly normal mutants	*Fabp4*	Hotamisligil et al. (1996)
7.A	Mutation in an X-linked gene	*Ar*	Yeh et al. (2002)
	Mutation in a Y-linked gene	*Ddx3y*	Rohozinski et al. (2002)
	Mutation in an imprinted autosomal gene	*Igf2*	DeChiara et al. (1991)
7.C	Dominant effects in chimeras	*Zic3*	Purandare et al. (2002)
7.D.1	Dominant effects in heterozygous mice	*Tbx1*	Jerome and Papaioannou (2001)
	Perinatal dominant lethal	*Sox9*	Bi et al. (2001)
7.D.2	Dominant reproductive problems	*Klhl10*	Yan et al. (2004)
8.A.1	Using chimeras to analyze a phenotype	*Tbx6*	Chapman et al. (2003)
Box 8.2	Two different alleles for genotyping chimeras	*Gsc*	Rivera-Perez et al. (1995)
		Tbx4	Naiche and Papaioannou (2003)
8.A.2.b	Tetraploid rescue of placental phenotype	*Ascl2*	Guillemot et al. (1994)
	ES-cell–diploid embryo chimeras	*Nodal*	Varlet et al. (1997)
	Chimeras made with trophoblast vesicles	*A*	Papaioannou and Gardner (1979)
	ES-cell–tetraploid embryo chimeras	*Foxa2*	Dufort et al. (1998)
	Chimera analysis with a tissue-specific reporter	*Egr2*	Voiculescu et al. (2001)
8.B	Transgene rescue of a mutant phenotype	*Myo15*	Probst et al. (1998)
8.C.1	Testing a conditional allele	*Lhx1*	Kwan and Behringer (2002)
	Floxed allele that is a hypomorph but can still be used as a conditional allele	*Bmp4*	Kulessa and Hogan (2002)
		Nodal	Lowe et al. (2001)
8.C.2	Incomplete excision by Cre resulting in hypomorphic phenotype	*Nr5a1*	Zhao et al. (2001)
	Conditional and tissue-specific mutation	*Fgf8*	Meyers et al. (1998)
	Inducible recombination	CreER™	Hayashi et al. (2001)
	Cre reporter		Soriano (1999)
8.D	Regulatory mutations	*Amh*	Arango et al. (1999)

References

Abuin A. and Bradley A. 1996. Recycling selectable markers in mouse embryonic stem cells. *Mol. Cell. Biol.* **16:** 1851–1856.

Akiyama H., Chaboissier M.C., Martin J.F., Schedl A., and de Crombrugghe B. 2002. The transcription factor Sox9 has essential roles in successive steps of the chondrocyte differentiation pathway and is required for expression of Sox5 and Sox6. *Genes Dev.* **16:** 2813–2828.

Arango N.A., Lovell-Badge R., and Behringer R.R. 1999. Targeted mutagenesis of the endogenous mouse *Mis* gene promoter: In vivo definition of genetic pathways of vertebrate sexual development. *Cell* **99:** 409–419.

Babinet C., Richoux V., Guenet J.L., and Renard J.P. 1990. The DDK inbred strain as a model for the study of interactions between parental genomes and egg cytoplasm in mouse preimplantation development. *Development* (Suppl.) **1990:** 81–87.

Baum M.J., Brown J.J.G., Kica E., Rubin B.S., Johnson R.S., and Papaioannou V.E. 1994. Effect of a null mutation of the c-*fos* proto-oncogene on sexual behavior of male mice. *Biol. Reprod.* **50:** 1040–1048.

Behringer R.R., Finegold M.J., and Cate R.L. 1994. Mullerian-inhibiting substance function during mammalian sexual development. *Cell* **79:** 415–425.

Bei M., Kratochwil K., and Maas R.L. 2000. BMP4 rescues a non-cell-autonomous function of Msx1 in tooth development. *Development* **127:** 4711–4718.

Bi W., Huang W., Whitworth D.J., Deng J.M., Zhang Z., Behringer R.R., and de Crombrugghe B. 2001. Haploinsufficiency of Sox9 results in defective cartilage primordia and premature skeletal mineralization. *Proc. Natl. Acad. Sci.* **98:** 6698–6703.

Bishop C.E., Whitworth D.J., Qin Y., Agoulnik A.I., Agoulnik I.U., Harrison W.R., Behringer R.R., and Overbeek P.A.

2000. A transgenic insertion upstream of sox9 is associated with dominant XX sex reversal in the mouse. *Nat. Genet.* **26:** 490–494.

Bonnette S.G. and Hadsell D.L. 2001. Targeted disruption of the IGF-I receptor gene decreases cellular proliferation in mammary terminal end buds. *Endocrinology* **142:** 4937–4945.

Braun R.E., Lo D., Pinkert C.A., Widera G., Flavell R.A., Palmiter R.D., and Brinster R.L. 1990. Infertility in male transgenic mice: Disruption of sperm development by HSV-tk expression in postmeiotic germ cells. *Biol. Reprod.* **43:** 684–693.

Brinster R.L. and Zimmermann J.W. 1994. Spermatogenesis following male germ-cell transplantation. *Proc. Natl. Acad. Sci.* **91:** 11298–11302.

Carmeliet P., Ferreira V., Breier G., Pollefeyt S., Kieckens L., Gertsenstein M., Fahrig M., Vandenhoeck A., Harpal K., Eberhardt C., Declercq C., Pawling J., Moons L., Collen D., Risau W., and Nagy A. 1996. Abnormal blood vessel development and lethality in embryos lacking a single VEGF allele. *Nature* **380:** 435–439.

Chaboissier M.C., Kobayashi A., Vidal V.I., Lutzkendorf S., van de Kant H.J., Wegner M., de Rooij D.G., Behringer R.R., and Schedl A. 2004. Functional analysis of Sox8 and Sox9 during sex determination in the mouse. *Development* **131:** 1891–1901.

Chang H., Huylebroeck D., Verchueren K., Guo Q., Matzuk M.M., and Zwijsen A. 1999. Smad5 knockout mice die at mid-gestation due to multiple embryonic and extraembryonic defects. *Development* **126:** 1631–1642.

Chapman D.L. and Papaioannou V.E. 1998. Three neural tubes in mouse embryos with mutations in the T-box gene, *Tbx6*. *Nature* **391:** 695–697.

Chapman D.L., Cooper-Morgan A., Harrelson Z., and Papaioannou V.E. 2003. Critical role for *Tbx6* in mesoderm specification in the mouse embryo. *Mech. Dev.* **120:** 837–847.

Chavez-Reyes A., Parant J.M., Amelse L.L., de Oca Luna R.M., Korsmeyer S.J., and Lozano G. 2003. Switching mechanisms of cell death in mdm2- and mdm4-null mice by deletion of p53 downstream targets. *Cancer Res.* **63:** 8664–8669.

Chen C.M. and Behringer R. 2004. Ovca1 regulates cell proliferation, embryonic development, and tumorigenesis. *Genes Dev.* **18:** 320–332.

Chen Z.F. and Behringer R.R. 1995. *twist* is required in head mesenchyme for cranial neural tube morphogenesis. *Genes Dev.* **9:** 686–699.

Chen Z.F., Rebelo S., White F., Malmberg A.B., Baba H., Lima D., Woolf C.J., Basbaum A.I., and Anderson D.J. 2001. The paired homeodomain protein DRG11 is required for the projection of cutaneous sensory afferent fibers to the dorsal spinal cord. *Neuron* **31:** 59–73.

Davenport T.G., Jerome-Majewska L.A., and Papaioannou V.E. 2003. Mammary gland, limb, and yolk sac defects in mice lacking *Tbx3*, the gene mutated in human ulnar mammary syndrome. *Development* **130:** 2263–2273.

Davis A.C., Wims M., Spotts G.D., Hann S.R., and Bradley A. 1993. A null c-myc mutation causes lethality before 10.5 days of gestation in homozygotes and reduced fertility in heterozygous female mice. *Genes Dev.* **7:** 671–682.

DeChiara T.M., Robertson E.J., and Efstratiadis A. 1991. Parental imprinting of the mouse insulin-like growth factor II gene. *Cell* **64:** 849–859.

de Rozieres S., Maya R., Oren M., and Lozano G. 2000. The loss of mdm2 induces p53-mediated apoptosis. *Oncogene* **19:** 1691–1697.

Donehower L.A., Harvey M., Slagle B.L., McArthur M.J., Montgomery C.A., Jr., Butel J.S., and Bradley A. 1992. Mice deficient for p53 are developmentally normal but susceptible to spontaneous tumours. *Nature* **356:** 215–221.

Dong S., Leung K.K., Pelling A.L., Lee P.Y., Tang A.S., Heng H.H., Tsui L.C., Tease C., Fisher G., Steel K.P., and Cheah K.S. 2002. Circling, deafness, and yellow coat displayed by yellow submarine (ysb) and light coat and circling (lcc) mice with mutations on chromosome 3. *Genomics* **79:** 777–784.

Downs K.M., Temkin R., Gifford S., and HcHugh J. 2000. Study of the murine allantois by allantoic explants. *Dev. Biol.* **233:** 347–364.

Dudas M., Nagy A., Laping N.J., Moustakas A., and Kaartinen V. 2004. Tgf-beta3-induced palatal fusion is mediated by Alk-5/Smad pathway. *Dev. Biol.* **266:** 96–108.

Dufort D., Schwartz L., Harpal K., and Rossant J. 1998. The transcription factor HNF3beta is required in visceral endoderm for normal primitive streak morphogenesis. *Development* **125:** 3015–3025.

Erickson J.T., Conover J.C., Borday V., Champagnat J., Barbacid M., Yancopoulos G., and Katz D.M. 1996. Mice lacking brain-derived neurotrophic factor exhibit visceral sensory neuron losses distinct from mice lacking NT4 and display a severe developmental deficit in control of breathing. *J. Neurosci.* **16:** 5361–5371.

Feldman B., Poueymirou W., Papaioannou V.E., DeChiara T.M., and Goldfarb M. 1995. Requirement of FGF-4 for postimplantation mouse development. *Science* **267:** 246–249.

Fitch K.R., McGowan K.A., van Raamsdonk C.D., Fuchs H., Lee D., Puech A., Herault Y., Threadgill D.W., Hrabe de Angelis M., and Barsh G.S. 2003. Genetics of dark skin in mice. *Genes Dev.* **17:** 214–228.

Friedrich G. and Soriano P. 1991. Promoter traps in embyonic stem cells: AA genetic screen to identify and mutate developmental genes in mice. *Genes Dev.* **5:** 1513–1523.

Furuta Y. and Hogan B.L. 1998. BMP4 is essential for lens induction in the mouse embryo. *Genes Dev.* **12:** 3764–3775.

Giese K.P., Martini R., Lemke G., Soriano P., and Schachner M. 1992. Mouse P0 gene disruption leads to hypomyelination, abnormal expression of recognition molecules, and degeneration of myelin and axons. *Cell* **71:** 565–576.

Gnarra J.R., Ward J.M., Porter F.D., Wagner J.R., Devor D.E., Grinberg A., Emmert-Buck M.R., Westphal H., Klausner R.D., and Linehan W.M. 1997. Defective placental vasculogenesis causes embryonic lethality in VHL-deficient mice. *Proc. Natl. Acad. Sci.* **94:** 102–107.

Goldin S.N. and Papaioannou V.E. 2003. Paracrine action of FGF4 during periimplantation development maintains trophectoderm and primitive endoderm. *Genesis* **36:** 40–47.

Gossler A., Joyner A.L., Rossant J., and Skarnes W.C. 1989. Mouse embryonic stem cells and reporter constructs to detect developmentally regulated genes. *Science* **244:** 463–465.

Guidi C.J., Sands A.T., Zambrowicz B.P., Turner T.K., Demers D.A., Webster W., Smith T.W., Imbalzano A.N., and Jones S.N. 2001. Disruption of Ini1 leads to peri-implantation lethality and tumorigenesis in mice. *Mol. Cell Biol.* **21:** 3598–3603.

Guillemot F., Nagy A., Auerbach A., Rossant J., and Joyner A.L. 1994. Essential role of *Mash-2* in extraembryonic development. *Nature* **371:** 333–336.

Hancock S.N., Agulnik S.I., Silver L.M., and Papaioannou V.E. 1999. Mapping and expression analysis the mouse ortholog of *Xenopus Eomesodermin*. *Mech. Dev.* **81:** 205–208.

Harrelson Z., Kelly R.G., Goldin S.N., Gibson-Brown J.J., Bollag R.J., Silver L.M., and Papaioannou V.E. 2004. *Tbx2* is essential for patterning the atrioventricular canal and for morphogenesis of the outflow tract during heart development. *Development* (in press).

Hasty P., Bradley A., Morris J.H., Edmondson D.G., Venuti J.M., Olson E.N., and Klein W.H. 1993. Muscle deficiency and neonatal death in mice with a targeted mutation in the myogenin gene. *Nature* **364:** 501–506.

Hayashi S., Tenzen T., and McMahon A.P. 2003. Maternal inheritance of Cre activity in a Sox2Cre deleter strain. *Genesis* **37:** 51–53.

Hayashi S., Lewis P., Pevney L., and McMahon A.P. 2002. Efficient gene modulation in mouse epiblast using a Sox2Cre transgenic mouse strain. *Mech. Dev.* (Suppl. 1) **119:** S97–S101.

Herceg Z., Hulla W., Gell D., Cuenin C., Lleonart M., Jackson S., and Wang Z.Q. 2001. Disruption of *Trrap* causes early embryonic lethality and defects in cell cycle progression. *Nat. Genet.* **29:** 206–211.

Heuberger B., Fitzka I., Wasner G., and Kratochwil K. 1982. Induction of androgen receptor formation by epithelium-mesenchyme interaction in embryonic mouse mammary gland. *Proc. Natl. Acad. Sci.* **79:** 2957–2961.

Hildebrand J.D. and Soriano P. 1999. Shroom, a PDZ domain-containing actin-binding protein, is required for neural tube morphogenesis in mice. *Cell* **99:** 485–497.

Holdener B.C. and Magnuson T. 1994. A mouse model for human hereditary tyrosinemia I. *BioEssays* **16:** 85–87.

Honrado G.I., Johnson R.S., Golombek D.A., Spiegelman B.M., Papaioannou V.E., and Ralph M.R. 1996. The circadian system of c-*fos* deficient mice. *J. Comp. Physiol. A* **178:** 563–570.

Horai R., Saijo S., Tanioka H., Nakae S., Sudo K., Okahara A., Ikuse T., Asano M., and Iwakura Y. 2000. Development of chronic inflammatory arthropathy resembling rheumatoid arthritis in interleukin 1 receptor antagonist-deficient mice. *J. Exp. Med.* **191:** 313–320.

Horie K., Yusa K., Yae K., Odajima J., Fischer S.E., Keng V.W., Hayakawa T., Mizuno S., Kondoh G., Ijiri T., Matsuda Y., Plasterk R.H., and Takeda J. 2003. Characterization of Sleeping Beauty transposition and its application to genetic screening in mice. *Mol. Cell Biol.* **23:** 9189–9207.

Hotamisligil G.S., Johnson R.S., Distel R.J., Ellis R., Papaioannou V.E., and Spiegelman B.M. 1996. Uncoupling of obesity from insulin resistance through a targeted mutation in aP2, the adipocyte fatty acid binding protein. *Science* **274:** 1377–1379.

Hrabe de Angelis M.H., Flaswinkel H., Fuchs H., Rathkolb B., Soewarto D., Marschall S., Heffner S., Pargent W., Wuensch K., Jung M., Reis A., Richter T., Alessandrini F., Jakob T., Fuchs E., Kolb H., Kremmer E., Schaeble K., Rollinski B., Roscher A., Peters C., Meitinger T., Strom T., Steckler T., Holsboer F., Klopstock T., Gekeler F., Schindelwolf C., Jung T., Avraham K., Behrendt H., Ring J., Zimmer A., Schughart K., Pfeffer K., Wolf E., and Balling R. 2000. Genome-wide, large-scale production of mutant mice by ENU mutagenesis. *Nat. Genet.* **25:** 444–447.

Jerome L.A. and Papaioannou V.E. 2001. DiGeorge syndrome phenotype in mice mutant for the T-box gene, *Tbx1*. *Nat. Genet.* **27:** 286–291.

Johnson R.S., Spiegelman B.M., and Papaioannou V.E. 1992. Pleiotropic effects of a null mutation in the c-*fos* proto-oncogene. *Cell* **71:** 577–586.

Karlsson L., Bondjers C., and Betsholtz C. 1999. Roles for PDGF-A and sonic hedgehog in development of mesenchymal components of the hair follicle. *Development* **126:** 2611–2621.

Kasarskis A., Manova K., and Anderson K.V. 1998. A phenotype-based screen for embryonic lethal mutations in the mouse. *Proc. Natl. Acad. Sci.* **95:** 7485–7490.

Kaul A., Koster M., Neuhaus H., and Braun T. 2000. Myf-5 revisited: Loss of early myotome formation does not lead to a rib phenotype in homozygous Myf-5 mutant mice. *Cell* **102:** 17–19.

Kelly M.A., Low M.J., Phillips T.J., Wakeland E.K., and Yanagisawa M. 2003. The mapping of quantitative trait loci underlying strain differences in locomotor activity between 129S6 and C57BL/6J mice. *Mamm. Genome* **14:** 692–702.

Kile B.T., Hentges K.E., Clark A.T., Nakamura H., Salinger A.P., Liu B., Box N., Stockton D.W., Johnson R.L., Behringer R.R., Bradley A., and Justice M.J. 2003. Functional genetic analysis of mouse chromosome 11. *Nature* **425:** 81–86.

Kruithof B.P., van den Hoff M.J., Wessels A., and Moorman A.F. 2003. Cardiac muscle cell formation after development of the linear heart tube. *Dev. Dynamics* **227:** 1–13.

Kuehn M.R., Bradley A., Robertson E.J., and Evans M.J. 1987. A potential animal model for Lesch-Nyhan syndrome through introduction of HPRT mutations into mice. *Nature* **326:** 295–298.

Kulessa H. and Hogan B.L. 2002. Generation of a loxP flanked bmp4loxP-lacZ allele marked by conditional lacZ expression. *Genesis* **32:** 66–68.

Kwan K.M. and Behringer R.R. 2002. Conditional inactivation of Lim1 function. *Genesis* **32:** 118–120.

Lerch-Gaggl A., Haque J., Li J., Ning G., Traktman P., and Duncan S.A. 2002. Pescadillo is essential for nucleolar assembly, ribosome biogenesis, and mammalian cell proliferation. *J. Biol. Chem.* **277:** 45347–45355.

Lettice L.A., Heaney S.J., Purdie L.A., Li L., de Beer P., Oostra B.A., Good D., Elgar G., Hill R.E., and de Graaf E. 2003. A long-range Shh enhancer regulates expression in the developing limb and fin and is associated with preaxial polydactyly. *Hum. Mol. Genet.* **12:** 1725–1735.

Li Y. and Behringer R.R. 1998. Esx1 is an X-chromosome-imprinted regulator of placental development and fetal growth. *Nat. Genet.* **20:** 309–311.

Lin Q., Schwarz J., Bucana C., and Olson E.N. 1997. Control of mouse cardiac morphogenesis and myogenesis by transcription factor MEF2C. *Science* **276:** 1404–1407.

Liu L., Czerwiec E., and Keefe D.L. 2004. Effect of ploidy and parental genome composition on expression of Oct-4 protein in mouse embryos. *Gene Expr. Patterns* **4:** 433–441.

Liu P., Wakamiya M., Shea M.J., Albrecht U., Behringer R.R., and Bradley A. 1999. Requirement for Wnt3 in vertebrate axis formation. *Nat. Genet.* **22:** 361–365.

Lowe L.A., Yamada S., and Kuehn M.R. 2001. Genetic dissection of *nodal* function in patterning the mouse embryo. *Development* **128:** 1831–1843.

Ludwig T., Chapman D.L., Papaioannou V.E., and Efstratiadis A. 1997. Targeted mutations of breast cancer susceptibility gene homologs in mice: Lethal phenotypes of *Brca1, Brca2, Brca1/Brca2, Brca1/p53*, and *Brca2/p53* nullizygous embryos. *Genes Dev.* **11:** 1226–1241.

Lufkin T., Dierich A., LeMeur M., Mark M., and Chambon P. 1991. Disruption of the Hox-1.6 homeobox gene results in defects in a region corresponding to its rostral domain of expression. *Cell* **66:** 1105–1119.

MacGregor G.R., Russell L.D., Van Beek M.E., Hanten G.R., Kovac M.J., Kozak C.A., Meistrich M.L., and Overbeek P.A. 1990. Symplastic spermatids (sys): A recessive insertional mutation in mice causing a defect in spermatogenesis. *Proc. Natl. Acad. Sci.* **87:** 5016–5020.

Meyers E.N., Lewandoski M., and Martin G.R. 1998. An *Fgf8* mutant allelic series generated by Cre- and Flp-mediated recombination. *Nat. Genet.* **18:** 136–141.

Miquerol L., Langille B.L., and Nagy A. 2000. Embryonic development is disrupted by modest increases in vascular endothelial growth factor gene expression. *Development* **127:** 3941–3946.

Mishina Y., Rey R., Finegold M.J., Matzuk M.M., Josso N., Cate R.L., and Behringer R.R. 1996. Genetic analysis of the Mullerian-inhibiting substance signal transduction pathway in mammalian sexual differentiation. *Genes Dev.* **10:** 2577–2587.

Miura T. and Shiota K. 2000. Time-lapse observation of branching morphogenesis of the lung bud epithelium in mesenchyme-free culture and its relationship with the localization of actin filaments. *Int. J. Dev. Biol.* **44:** 899–902.

Montes de Oca Luna R., Wagner D.S., and Lozano G. 1995. Rescue of early embryonic lethality in mdm2-deficient mice by deletion of p53. *Nature* **378:** 203–206.

Morasso M.I., Grinberg A., Robinson G., Sargent T.D., and Mahon K.A. 1999. Placental failure in mice lacking the homeobox gene Dlx3. *Proc. Natl. Acad. Sci.* **96:** 162–167.

Morgan D., Turnpenny L., Goodship J., Dai W., Majumder K., Matthews L., Gardner A., Schuster G., Vien L., Harrison W., Elder F.F., Penman-Splitt M., Overbeek P., and Strachan T. 1998. Inversin, a novel gene in the vertebrate left-right axis pathway, is partially deleted in the *inv* mouse. *Nat. Genet.* **20:** 149–156.

Naiche L.A. and Papaioannou V.E. 2003. Loss of *Tbx4* blocks hindlimb development and affects vascularization and fusion of the allantois. *Development* **130:** 2681–2693.

Nakashima K., Zhou X., Kunkel G., Zhang Z., Deng J.M., Behringer R.R., and de Crombrugghe B. 2002. The novel zinc finger-containing transcription factor osterix is required for osteoblast differentiation and bone formation. *Cell* **108:** 17–29.

Nichols J., Chambers I., Taga T., and Smith A. 2001. Physiological rationale for responsiveness of mouse embryonic stem cells to gp130 cytokines. *Development* **128:** 2333–2339.

Ohsugi M., Larue L., Schwarz H., and Kemler R. 1997. Cell-junctional and cytoskeletal organization in mouse blastocysts lacking E-cadherin. *Dev. Biol.* **185:** 261–271.

Ohtoshi A., Wang S.W., Maeda H., Saszik S.M., Frishman L.J., Klein W.H., and Behringer R.R. 2004. Regulation of retinal cone bipolar cell differentiation and photopic vision by the CVC homeobox gene Vsx1. *Curr. Biol.* **14:** 530–536.

Palmiter R.D., Brinster R.L., Hammer R.E., Trumbauer M.E., Rosenfeld M.G., Birnberg N.C., and Evans R.M. 1982. Dramatic growth of mice that develop from eggs microinjected with metallothionein-growth hormone fusion genes. *Nature* **300:** 611–615.

Papaioannou V.E. and Gardner R.L. 1979. Investigation of the lethal yellow Ay/Ay embryo using mouse chimaeras. *J. Embryol. Exp. Morphol.* **52:** 153–163.

Paylor R., Johnson R.S., Papaioannou V.E., Speigelman B.M., and Wehner J.M. 1994. Behavioral assessment of c-*fos* mutant mice. *Brain Res.* **651:** 275–282.

Perea-Gomez A., Vella F.D., Shawlot W., Oulad-Abdelghani M., Chazaud C., Meno C., Pfister V., Chen L., Robertson E., Hamada H., Behringer R.R., and Ang S.L. 2002. Nodal antagonists in the anterior visceral endoderm prevent the formation of multiple primitive streaks. *Dev. Cell* **3:** 745–756.

Pevny L., Simon M.C., Robertson E., Klein W.H., Tsai S.F., D'Agati V., Orkin S.H., and Costantini F. 1991. Erythroid differentiation in chimaeric mice blocked by a targeted mutation in the gene for transcription factor GATA-1. *Nature* **349:** 257–260.

Probst F.J., Fridell R.A, Raphael Y., Saunders T.L., Wang A., Liang Y., Morell R.J., Touchman J.W., Lyons R.H., Noben-Trauth K., Friedman T.B., and Camper S.A. 1998. Correction of deafness in shaker-2 mice by an unconventional myosin in a BAC transgene. *Science* **280:** 1444–1447.

Purandare S.M., Ware S.M., Kwan K.M., Gebbia M., Bassi M.T., Deng J.M., Vogel H., Behringer R.R., Belmont J.W., and Casey B. 2002. A complex syndrome of left-right axis, central nervous system and axial skeleton defects in Zic3 mutant mice. *Development* **129:** 2293–2302.

Rinkenberger J.L., Horning S., Klocke B., Roth K., and Korsmeyer S.J. 2000. Mcl-1 deficiency results in peri-implantation embryonic lethality. *Genes Dev.* **14:** 23–27.

Rivera-Perez J.A., Mallo M., Gendron-Maguire M., Gridley T., and Behringer R.R. 1995. Goosecoid is not an essential component of the mouse gastrula organizer but is required for craniofacial and rib development. *Development* **121:** 3005–3012.

Roffler-Tarlov S., Gibson Brown J.J., Tarlov E., Stolarov J., Chapman D.L., Alexiou M., and Papaioannou V.E. 1996. Programmed cell death in the absence of c-Fos and c-Jun. *Development* **122:** 1–9.

Rohozinski J., Agoulnik A.I., Boettger-Tong H.L., and Bishop C.E. 2002. Successful targeting of mouse Y chromosome genes using a site-directed insertion vector. *Genesis* **32:** 1–7.

Rudnicki M.A., Schnegelsberg P.N., Stead R.H., Braun T., Arnold H.H., and Jaenisch R. 1993. MyoD or Myf-5 is required for the formation of skeletal muscle. *Cell* **75:** 1351–1359.

Satokata I. and Maas R. 1994. Msx1 deficient mice exhibit cleft palate and abnormalities of craniofacial and tooth development. *Nat. Genet.* **6:** 348–356.

Schlatt S., Honaramooz A., Boiani M., Scholer H.R., and Dobrinski I. 2003. Progeny from sperm obtained after ectopic grafting of neonatal mouse testes. *Biol. Reprod.* **68:** 2331–2335.

Sekine K., Ohuchi H., Fujiwara M., Yamasaki M., Yoshizawa T., Sato T., Yagishita N., Matsui D., Koga Y., Itoh N., and Kato S. 1999. Fgf10 is essential for limb and lung formation. *Nat. Genet.* **21:** 138–141.

Shawlot W. and Behringer R.R. 1995. Requirement for Lim1 in head-organizer function. *Nature* **374:** 425–430.

Shi C.Z., Collins H.W., Buettger C.W., Garside W.T., Matschinsky F.M., and Heyner S. 1994. Insulin family growth factors have specific effects on protein synthesis in preimplantation mouse embryos. *Mol. Reprod. Dev.* **37:** 398–406.

Shiojiri N., Wada J.I., Tanaka T., Noguchi M., Ito M., and Gebhardt R. 1995. Heterogeneous hepatocellular expression of glutamine synthetase in developing mouse liver and in testicular transplants of fetal liver. *Lab. Invest.* **72:** 740–747.

Soriano P. 1997. The PDGF alpha receptor is required for neural crest cell development and for normal patterning of the somites. *Development* **124:** 2691–2700.

Soriano P. 1999. Generalized *lacZ* expression with the *ROSA26* Cre reporter. *Nat. Genet.* **21:** 70–71.

Soriano P., Montgomery C., Geske R., and Bradley A. 1991. Targeted disruption of the c-src proto-oncogene leads to osteopetrosis in mice. *Cell* **64:** 693–702.

Srinivas S., Goldberg M.R., Watanabe T., D'Agati V., al-Awqati Q., and Costantini F. 1999. Expression of green fluorescent protein in the ureteric bud of transgenic mice: A new tool for the analysis of ureteric bud morphogenesis. *Dev. Genet.* **24:** 241–251.

Stewart C.L., Kaspar P., Brunet L.J., Bhatt H., Gadi I., Kontgen F., and Abbondanzo S.J. 1992. Blastocyst implantation depends on maternal expression of leukaemia inhibitory factor. *Nature* **359:** 76–79.

Threadgill D.W., Dlugosz A.A., Hansen L.A., Tennenbaum T., Lichti U., Yee D., LaMantia C., Mourton T., Herrup K., Harris R.C., et al. 1995. Targeted disruption of mouse EGF receptor: Effect of genetic background on mutant phenotype. *Science* **269:** 230–234.

van Deursen J., Boer J., Kasper L., and Grosveld G. 1996. G2 arrest and impaired nucleocytoplasmic transport in mouse embryos lacking the proto-oncogene CAN/Nup214. *EMBO J.* **15:** 5574–5583.

Varlet I., Collignon J., and Robertson E.J. 1997. Nodal expression in the primitive endoderm is required for specification of the anterior axis during mouse gastrulation. *Development* **124:** 1033–1044.

Voiculescu O., Taillebourg E., Pujades C., Kress C., Buart S., Charnay P., and Schneider-Maunoury S. 2001. Hindbrain patterning: Krox20 couples segmentation and specification of regional identity. *Development* **128:** 4967–4978.

Watabe-Rudolph M., Schlautmann N., Papaioannou V.E., and Gossler A. 2002. The mouse rib-vertebrae mutation is a hypomorphic *Tbx6* allele. *Mech. Dev.* **119:** 251–256.

Winnier G., Hargett L., and Hogan B.L. 1997. The winged helix transcription factor MFH1 is required for proliferation and patterning of paraxial mesoderm in the mouse embryo. *Genes Dev.* **11:** 926–940.

Xue Y., Gao X., Lindsell C.E., Norton C.R., Chang B., Hicks C., Grendron-Maguire M., Rand E.B., Weinmaster G., and Gridley T. 1999. Embryonic lethality and vascular defects in mice lacking the Notch ligand Jagged1. *Hum. Mol. Genet.* **8:** 723–730.

Yan W., Ma L., Burns K.H., and Matzuk M.M. 2004. Haploinsufficiency of kelch-like protein homolog 10 causes infertility in male mice. *Proc. Natl. Acad. Sci.* **101:** 7793–7798.

Yeh S., Tsai M.Y., Xu Q., Mu X.M., Lardy H., Huang K.E., Lin H., Yeh S.D., Altuwaijri S., Zhou X., Xing L., Boyce B.F., Hung M.C., Zhang S., Gan L., Chang C., and Hung M.C. 2002. Generation and characterization of androgen receptor knockout (ARKO) mice: An in vivo model for the study of androgen functions in selective tissues. *Proc. Natl. Acad. Sci.* **99:** 13498–13503.

Yoder M.C., Papaioannou V.E., Breitfeld P.P., and Williams D.A. 1994. Murine yolk sac endoderm- and mesoderm-derived cell lines support in vitro growth and differentiation of hematopoietic cells. *Blood* **83:** 2436–2443.

Zeitlin S., Liu J.-P., Chapman D.L., Papaioannou V.E., and Efstratiadis A. 1995. Increased apoptosis and early embryonic lethality in mice nullizygous for the Huntington's disease gene homologue. *Nat. Genet.* **11:** 155–163.

Zhao L., Bakke M., Krimkevich Y., Cushman L.J., Parlow A.F., Camper S.A., and Parker K.L. 2001. Hypomorphic phenotype in mice with pituitary-specific knockout of steroidogenic factor 1. *Genesis* **30:** 65–69.

Zhao M., Shirley C.R., Hayashi S., Marcon L., Mohapatra B., Suganuma R., Behringer R.R., Boissonneault G., Yanagimachi R., and Meistrich M.L. 2004. Transition nuclear proteins are required for normal chromatin condensation and functional sperm development. *Genesis* **38:** 200–213.

Zhao Q., Behringer R.R., and de Crombrugghe B. 1996. Prenatal folic acid treatment suppresses acrania and meroanencephaly in mice mutant for the *Cart1* homeobox gene. *Nat. Genet.* **13:** 275–283.

Appendix 3

Glossary

AGM, aorta-gonad-mesonephros, a blood-cell-forming region of the embryo.

AI, artificial insemination, experimental introduction of a viable sperm preparation into the female reproductive tract, usually delivered directly through the cervix. This procedure can be used to overcome infertility.

allantois, a mesoderm-derived extraembryonic structure that forms at the posterior pole of the embryo, grows through the extraembryonic coelom to attach to the chorion, and differentiates into the umbilical cord.

alleles, alternative or variant forms of the same gene occurring at the same locus on homologous chromosomes. Except in cases of duplication or deletion, an animal carries two alleles of autosomal genes, one derived from each parent.

allelic series, more than two different alleles of a given gene or locus. The phenotypic effects of an allelic series can sometimes be arranged in a dominance hierarchy with graded phenotypic effects.

alternative splicing, joining of different or alternative exons from primary transcripts from a single gene to produce different mRNA and protein isoforms.

anemia, the condition of a reduced number of circulating red blood cells relative to wild type.

aneuploid, a deviation from the standard karyotype, which in the mouse consists of 20 pairs of chromosomes including 19 autosomal pairs and the X and Y sex chromosomes. Such deviations can involve duplications, deletions, or translocations of whole or partial chromosomes.

anlagen, tissue primordium.

antimorph, a dominant-negative mutation encoding a mutant protein that interferes with the function of the wild-type protein.

autosomal, pertaining to chromosomes other than the X and Y sex chromosomes.

BAC, bacterial artificial chromosome.

backcross, breeding a mouse to one of its parental strains. This approach is used with forced heterozygosity to transfer an allele onto a different strain background by repeated backcrossing.

background genotype, *see genetic background.*

balancer chromosome, a chromosome with an inversion marked by a dominant visible marker (e.g., coat color). The inversion may or may not have a linked recessive lethal mutation. Balancer chromosomes can be created by chromosome engineering and are used in mutagenesis screens and for maintaining recessive lethal or sterile mutations.

β-gal, beta-galactosidase, protein product of the *E. coli lacZ* gene.

bicistronic, an mRNA transcript encoding two different proteins; may be engineered using an internal ribosomal entry site.

blastocoel, the cavity of the blastocyst, also called the blastocyst cavity.

blastocyst, preimplantation-stage embryo between approximately 3.5 and 4.5 dpc, consisting of approximately 32–128 cells arranged as a single layer of trophoblast cells surrounding an inner cell mass, with an asymmetrically located cavity, the blastocyst cavity.

blastomeres, cells of early preimplantation embryos between the 2-cell and approximately 16- to 32-cell stage. They are the products of cleavage in which cell division is not accompanied by cell growth.

cell autonomous, activity of a gene (or protein) that has its effect on the cell in which it is produced.

Chi squared (χ^2), test of probability to assess whether significant differences exist between an observed number of events and an expected number based on the null hypothesis. This method is commonly used to test for deviations from expected Mendelian segregation ratios.

chimera, a name appropriated from a mythical creature with the head of a lion, the body of a goat, and the tail of a serpent, now used to describe experimental mice made by the combination of two or more embryos or the combination of embryonic stem cells with an embryo. Genetic differences among the components of the chimera provide a means of identifying the cellular progeny of each component.

chorioallantoic placenta, definitive placenta created by fusion of the allantois to the chorion.

chorion, an extraembryonic membrane in the early postimplantation embryo composed of extraembryonic ectoderm and mesoderm. The allantois fuses with this membrane to form part of the chorioallantoic placenta.

CL, corpus luteum (plural: **corpora lutea**), the ovarian progesterone-secreting organ that forms from a collapsed follicle following ovulation.

cleavage, the process of cell division during preimplantation development, when cell division is not accompanied by cell growth and thus the size of the cells, known as blastomeres, decreases.

CNS, central nervous system, comprised of the brain and spinal cord.

codominant, the nature of inheritance of phenotype. A codominant phenotype exists when the effects of both alleles can be simultaneously detected.

co-isogenic strains, stains that differ from one another at a single locus, usually through the occurrence of a spontaneous mutation in an inbred strain or through targeted mutagenesis in an inbred strain.

compaction, process whereby the blastomeres of the preimplantation embryo pack into a solid ball of cells, cell–cell adhesion increases, and blastomeres flatten onto one another as a result of increased cell–cell contacts.

complementation test, cross to determine whether two different mutations are within the same gene.

compound heterozygote, an animal carrying two different mutant alleles at a particular locus.

congenic strains, strains that differ from one another at a single locus and at closely linked loci as well. Congenic strains are produced by repeated backcrossing of mice carrying a mutation to an inbred strain for at least ten generations while maintaining heterozygosity for the mutation. Congenic strains differ from co-isogenic strains in that they are produced by breeding. They also differ by a small chromosomal segment, rather than by a single allele.

construct, DNA fragment(s) used to modify the mouse genome, e.g., a targeting construct.

corpora lutea, *see CL*.

Cre, DNA recombinase from *E. coli* bacteriophage P1 that recognizes the 34-bp *loxP* sequence.

C-section, Caesarian section, transabdominal delivery of fetal mice, usually done by sacrificing the mother.

cytospin, cell suspension centrifuged onto a microscope slide.

decidua, uterine tissue that differentiates around the implanting blastocyst. It is shed at parturition, hence the name.

decidualization, the reaction of the maternal uterine epithelium to an implanting embryo (or to certain other types of stimulus) that includes the development of decidual tissue surrounding the implantation site, some of which will take part in the formation of the definitive, chorioallantoic placenta.

deleter mice, genetically engineered mice that express Cre or Flp recombinase in the germ line or in the embryonic cells that give rise to the germ line. These mice are typically used to remove drug-selection cassettes.

DIC, differential interference contrast microscopy.

diploid, containing two homologous sets of chromosomes, referred to as 2N. Most somatic cells in mammals are diploid and divide by mitosis.

dominant, refers to the nature of inheritance of phenotypes and the relationship of one allele to a second at the same locus. A dominant phenotype is detectable when only one variant allele is present. A single mutation can be dominant for some phenotypic traits and recessive for others.

double heterozygote, an animal carrying one mutant allele at each of two different loci.

dpc, days postcoitus, the number of days after mating, used to indicate the age of embryos.

electrofusion, electrical pulse that disrupts plasma membranes between cells, leading to fusion of the two cells. This method is used to fuse blastomeres of two-cell-stage embryos to generate tetraploid embryos for chimera production.

embryo proper, the body of the fetus excluding fetal membranes (extraembryonic components) and placenta.

endochondral ossification, bone formation through a cartilage template.

endoreduplication, repeated rounds of DNA replication in the absence of intervening cell division. This process is characteristic of the trophoblastic giant cells and leads to polyploidy.

ENU, *N*-ethyl-*N*-nitrosourea, a chemical mutagen that usually causes single base-pair changes.

epiblast, tissue of the blastocyst that gives rise to the ectoderm, mesoderm, and endoderm of the embryo proper, as well as the extraembryonic mesoderm.

epistasis, the hierarchical relationship between two genes in a genetic pathway such that the effects of one mutation mask the effects of another.

erythrocyte, red blood cell.

ES cells, embryonic stem cells, pluripotent stem cells derived from the inner cell mass of blastocysts that can be perpetuated in vitro as undifferentiated cells, but that have a developmental potential similar to epiblast cells, i.e., in chimeras they can contribute to all tissues of the fetus and the extraembryonic mesoderm but not extraembryonic endoderm or trophectoderm.

EST, expressed sequence tag, partial sequence from a cDNA clone.

euploid, the standard karyotype that in mice consists of 20 pairs of chromosomes including 19 autosomal pairs and the X and Y sex chromosomes.

expressivity, the severity of a mutant phenotype within an individual. This can vary among individuals with the same mutation.

extraembryonic ectoderm, a trophoblast-derived layer in the early postimplantation embryo that gives rise to part of the placenta.

extraembryonic, embryo-derived tissues, such as the amnion, chorion, and yolk sac, that will not make any cellular contribution to the body of the fetus. The term refers to tissue that is not part of the embryo proper.

FACS, fluorescence-activated cell sorting.

floxed, flanked by *loxP* sites.

flrted, flanked by *FRT* sites.

forced heterozygosity, selection of heterozygous mice as breeding stock for the next generation. This approach is used in backcrossing to transfer an allele to a different strain background or during inbreeding to maintain a mutation in the stock.

gametes, the haploid sex cells, sperm (spermatozoa), or oocytes that are the products of meiotic divisions and come together during fertilization to form the zygote.

gametogenesis, the process of formation of the gametes that includes meiosis and the cellular differentiation of the mature sperm and oocyte.

gene targeting, homologous recombination between DNA sequences residing in the chromosome and introduced DNA sequences. This technology allows for directed mutagenesis of a specific locus.

gene trap, an insertional mutagen usually containing a splice acceptor sequence and reporter; chimeric cDNAs that can facilitate cloning are generated.

genetic background, usually referring to the strain or stock on which a mutant allele is maintained, e.g., an inbred C57BL/6J background or an outbred background.

genetic modifier, a locus that enhances or suppresses the expression of another locus.

genotype, the genetic makeup of an organism as opposed to the physical appearance or phenotype; the combination of alleles at a specific locus.

germ layers, the three classic layers of a triploblastic embryo: the ectoderm, mesoderm, and endoderm.

germ-line transmission, the inheritance of an ES-cell-derived allele of a chimera by its offspring.

gestation, period of embryogenesis, which takes place within the maternal reproductive tract.

GFP, green fluorescent protein. eGFP is enhanced green fluorescent, a variant with enhanced spectral qualities.

giant cells, trophoblast cells with very large nuclei resulting from endoreduplication of DNA. These cells surround the embryo at the maternal–embryo interface and form the outside fetal layer of the chorioallantoic placenta.

haploid, containing a single set of chromosomes, referred to as N. Haploid cells result from meiotic division.

haploinsufficiency, a phenotype resulting from the presence of half the normal amount of gene product due to, e.g., a null mutation in one allele, a deletion of one allele, imprinting of an allele, or X-chromosome inactivation. Contrast this phenotype with haplosufficiency, in which the amount of gene product generated by one allele is sufficient for normal function.

hatching, the process whereby the blastocyst escapes from the zona pellucida before implantation.

hematopoiesis, the process of blood cell formation.

hemizygous, a locus with no corresponding locus on the homologous chromosome. This term usually refers to an insertional transgene or X- and Y-chromosome-linked genes in males.

heterozygote, an animal that has two distinguishable alleles at a particular locus. The locus or the animal is considered heterozygous.

histocompatible, during transplantation, donor tissue is recognized by the recipient as self.

homolog/homologous, (1) member of a chromosome pair in diploid organisms; (2) genes in one species that have a common evolutionary origin and function as genes from another species, or genes within a species that have a common origin and function. (*See also ortholog.* Note that all orthologs are homologs but not all homologs are orthologs.)

homologous DNA, genetically identical DNA.

homologous recombination, genetic exchange between homologous DNA, e.g., between homologous chromosomes or between a chromosome and a gene-targeting construct containing homologous genomic DNA. (*See also gene targeting.*)

homozygote, an organism that has two identical alleles at a particular locus. The locus or animal is considered homozygous. A double homozygote is an animal that is homozygous at two particular loci.

hybrid mice, first filial (F_1) generation of mice from a cross between two inbred strains. These mice are all genetically uniform, i.e., identical to each other. However, when hybrid mice are mated together, the offspring, the F_2 generation, will be genetically nonuniform because these offspring will be segregating for all loci at which the parental strains differ; thus, each F_2 animal will have a unique, recombinant genetic makeup.

hybrid vigor, generally increased vigor and fecundity associated with heterozygosity at a large number of loci. This term is usually used to refer to the offspring of a cross between two inbred strains that will be more vigorous than either parent strain.

hypermorph, a mutation that causes an increase in gene expression or protein activity compared to wild type.

hypomorph, a mutation that retains some wild-type function of the gene and results in a less severe phenotype than a null, or loss-of-function, allele.

ICM, inner cell mass, the cells enclosed within the trophectoderm layer at the blastocyst stage of development.

ICSI, intracytoplasmic sperm injection, sperm or sperm heads injected directly into oocytes. This method can be used to overcome certain types of infertility.

immunosurgery, a procedure for isolation of the ICM of the blastocyst by exposure of the embryo to antibodies, followed by complement-mediated lysis of the outer trophoblast layer.

implantation delay, also called diapause, a physiological delay in the implantation process accompanied by a delay in embryonic development. In mice, implantation delay can be brought about by lactation and nursing of a previous litter, or experimentally by ovariectomy of a recently pregnant female.

implantation, the process whereby the blastocyst attaches to and invades the maternal uterine epithelium, elicits a decidual response in the mother, and establishes maternal–fetal exchange through placental structures. In the mouse, this process begins at 4.5 dpc.

imprinting, a process of epigenetic silencing of one allele of a homologous pair that takes place at certain loci during gametogenesis. Some loci are maternally imprinted and others are paternally imprinted.

inbred strains, for all practical purposes, the mice of an inbred strain are genetically identical to one another except for sex chromosome differences. Inbred strains are derived by many generations of close inbreeding, minimally 20 generations of brother-sister matings. However, if separate breeding colonies of inbred strains are maintained after the initial derivation, they are called substrains after 18 generations of separation in recognition of the probable accumulation of random genetic mutations that render them nonidentical.

inbreeding depression, general loss of vigor and fertility as a result of the loss of heterozygosity or hybrid vigor during the inbreeding process.

inbreeding, breeding scheme that results in increased homozygosity and is usually accomplished by brother-sister or father-daughter matings. A strain is considered inbred after 20 generations of brother-sister matings, when the probability of heterozygosity at any given locus is less than 2%. Inbreeding will occur in any small closed breeding colony.

insertional mutation, a mutation in an endogenous gene brought about by the insertion of a segment of DNA.

intercross, a cross between two animals that have the same heterozygous genotype at a specific locus.

IRES, internal ribosome entry site, used to create bicistronic mRNAs.

isogenic, genetically identical, as in DNA derived from different individuals of an inbred strain.

IVF, in vitro fertilization, combining sperm and oocytes in vitro for fertilization; usually followed by transfer of the resulting embryos to the oviduct or uterus for continued development.

karyotype, the number of chromosomes and the specific microscopic banding pattern seen upon staining of condensed chromatin.

knockout mice, jargon that originally referred to mice with a nonfunctional allele produced by gene targeting in ES cells, but which has been generalized for use with all mutations made by the method. Because of this imprecision, the term should be well defined or avoided altogether.

labyrinth, region of the placenta in which maternal and fetal tissues are tightly juxtaposed and nutrient and waste exchange takes place.

locus (plural: **loci**), the position or location of a gene on a chromosome.

loss of function, referring to a mutation in which the protein is absent or rendered nonfunctional; also called a null or null mutation.

loxP, locus of crossing-over, 34-bp sequence recognized by Cre recombinase. Contains two 13-bp inverted repeats and an 8-bp asymmetric spacer region in between.

marker recycling, method for creating homozygous mutant ES cell lines. A gene-targeting vector with a floxed drug-selection cassette is targeted to one allele and Cre recombinase is used to remove the drug-selection cassette. The heterozygous neo^- ES cells are then used to target the remaining wild-type allele, using the same gene-targeting vector and G418 selection.

MEFs, mouse embryonic fibroblasts.

meiosis, the nuclear reduction division process by which diploid germ cell precursors segregate chromosomes into haploid nuclei of sperm and oocytes; through this process, one diploid germ cell produces four haploid daughter cells.

mesonephros (plural: **mesonephroi**), the middle one of the three pairs of embryonic renal organs that develop in most vertebrates.

MI, mitotic index, a measure of cell division determined by dividing the number of cells in mitosis by the total number of cells examined.

moribund, near death.

morula, a preimplantation embryo of between 8 and 32 cells. The name derives from the mulberry appearance of the blastomeres before compaction. A compacted morula describes the postcompaction embryo before blastocyst formation.

mosaic, in general terms, an animal consisting of genetically different cells, regardless of how it originated. However, the term has been more specifically used to distinguish an animal, such as one with a somatic mutation that is derived from a single zygote, from a chimera, which is derived from two or more different zygotes.

mutant allele, a variant allele that usually confers a phenotypically identifiable difference with reference to a "wild-type" phenotype. The term usually, but not necessarily, refers to an allele with a deleterious effect.

mutation, a new allele present in an organism that is not present in its parents. Spontaneous mutations occur without experimental intervention, whereas induced mutations are brought about by experimental means such as chemical mutagenesis or homologous recombination.

***neo*,** neomycin resistance gene or neomycin resistance expression cassette. *neo* is a bacterial gene encoding aminoglycoside-3-phosphotransferase that confers resistance to the aminoglycoside G418.

neomorph, mutant protein with a new activity in comparison to wild type.

null, referring to a mutation that eliminates gene function.

OCT, Tissue-Tek® Optimal Cutting Temperature Compound, a liquid embedding compound to freeze tissues for frozen sectioning.

oogenesis, the process of oocyte formation from the female primordial germ cells; female gametogenesis.

oogonia, diploid female germ cells that give rise to oocytes.

ORF, open reading frame.

organogenesis, the process of organ formation.

ortholog/orthologous, genes in two species that are related by evolution from a single common ancestral gene. Several genes in one species may have a single ortholog in another species and vice versa. (*See also homolog.* Note that all orthologs are homologs but not all homologs are orthologs.)

orthotopic, in the normal or usual position.

outbred or noninbred, also called random bred, although the breeding scheme is not random but carefully controlled to avoid inbreeding. Outbred mice are usually bred in large closed colonies. These stocks are segregating for many alleles, some of which may be defined, and are generally robust mice. Different outbred stocks will vary in the range of alleles they contain.

outcross, a mating between genetically unrelated animals.

parietal yolk sac, extraembryonic membrane composed of trophoblast giant cells and parietal endoderm with the acellular Reichert's membrane; it lines the yolk sac cavity.

parturition, the process of birth.

PCNA, proliferating cell nuclear antigen, marks cells in mitosis and is used in an immunocytochemical assay for cell proliferation.

penetrance, percentage of genetically mutant individuals that exhibit the mutant phenotype. The mutant phenotype is fully penetrant when all mutant individuals exhibit the mutant phenotype.

pericardial effusion, edema resulting in a fluid-filled pericardial sac.

perinatal, time period soon after birth.

PFA, paraformaldehyde, a tissue fixative often used for embryos.

pgk, *phosphoglycerate kinase 1 gene.*

phenotype, the result of interaction between genotype and the environment. Like any measurable characteristic of an organism, it can be determined by any assay.

PNS, peripheral nervous system.

polymorphic, more than one allele for a specific gene or DNA segment.

postnatal, after birth.

pronucleus (plural: pronuclei), nuclear structure derived from either the sperm nucleus or the oocyte nucleus that forms in the zygote after fertilization.

provirus, viral DNA that has integrated into the host genome.

pseudopregnant, a physiological state brought about in an estrous female by the stimulus of mating. Pseudopregnancy parallels the hormonal stages of early pregnancy even though there are no embryos. It can be induced by mating with vasectomized males to provide females as recipients for embryo transfer.

RBCs, red blood cells, erythrocytes.

recessive, the nature of inheritance of phenotypes. In relation to another allele at the same locus, a recessive phenotype is one that is not detectable unless both alleles have a particular mutation. A single mutation can be dominant for some phenotypic traits and recessive for others.

redundancy, overlap in function of different genes such that loss of function of one gene product is insufficient to cause an overt mutant phenotype, but concomitant loss of function of the product of a second, related gene leads to a mutant phenotype.

rescuing transgene, a transgene randomly integrated into the genome that can provide wild-type function to suppress a mutant phenotype.

resorption or resorption site, the remains of a dead conceptus and its implantation site in the uterus.

retrovirus, RNA virus whose life cycle includes reverse transcription to generate DNA that integrates into the host genome as a provirus. This serves as a template to make more virus.

Rosa26, reverse orientation splice acceptor 26, a locus that was discovered in a gene trap experiment that has ubiquitous expression in the embryo proper, but patchy expression in the extraembryonic tissues.

semidominant, the nature of inheritance of phenotypes. A semidominant phenotype occurs when the phenotype in the heterozygote is intermediate between the wild-type and homozygous mutant.

speed congenic, in transferring a mutation onto an inbred strain, this method makes use of genetic markers to select animals for breeding in the next generation that have undergone specific chromosomal recombination events, thus increasing the genetic contribution of the inbred strain.

spermatogenesis, the process of gametogenesis in the male germ line resulting in the production of spermatozoa.

spermatogonia, diploid male germ cells that give rise to spermatozoa.

spermiogenesis, the stage of spermatogenesis in which haploid spermatids are morphologically transformed into spermatozoa.

SPF, specific pathogen free, a categorization of animal facilities maintained to exclude a specific group of bacterial and viral pathogens.

spongiotrophoblast, the layer of the chorioallantoic placenta, derived from the trophoblast and surrounded by trophoblast giant cells.

stock, referring to outbred mice.

substrain, an inbred strain maintained as a separate breeding population from the parent strain for 18 generations or more. Substrains may differ because of the fixation of spontaneous mutations or because of genetic contamination.

superovulation, hormonal induction of ovulation of a large number of oocytes using timed injections of pregnant mare serum gonadotropin and human chorionic gonadotropin.

syncytiotrophoblast, a syncytium of multinucleated trophoblast cells that makes up the labyrinth of the placenta.

syntenic, loci in the same linkage group. Conserved synteny between species refers to the situation in which linked loci in one species have orthologs that are linked in another species.

targeted mutagenesis, the technology that allows for directed mutagenesis of a specific locus

TEM, transmission electron microscopy.

teratocarcinoma, malignant tumor similar in origin to a teratoma, but one in which a permanent population of stem cells renders the tumor transplantable. Embryonal carcinoma cells are derived from teratocarcinomas.

teratoma, a benign tumor with a variety of differentiated cell types. Teratomas are derived from the transplantation of embryos, genital ridges, or other fragments of embryos. The range of differentiated cell types is a reflection of the developmental potential of the transplanted tissue.

term or **full term**, full duration of gestation.

tetraploid, containing two diploid sets of chromosomes, or 4N. Tetraploids are typically induced by electrofusion of blastomeres at the two-cell stage.

totipotency, the potential to differentiate into all cell types including embryonic and extraembryonic tissues.

transgene, foreign DNA that has been introduced and incorporated into the genome.

transgenic mice, mice containing foreign DNA that has been introduced into their genome. Technically, this includes mice made by viral infection or injection of DNA into preimplantation embryos (which results in random insertion of the foreign DNA) as well as mice produced by targeted mutagenesis via homologous recombination in ES cells. However, the term is usually reserved for mice made by random insertion of DNA following zygote injection.

transposase, enzyme that can excise and insert a transposon into DNA.

transposon, mobile DNA element.

trophectoderm, the outermost layer of epithelial cells of the blastocyst of the preimplantation embryo; also called trophoblast.

trophoblast giant cells, the nondividing, polyploid cells of the chorioallantoic placenta that are formed by endoreduplication and derived from the trophectoderm layer of the blastocyst.

trophoblast, the outer layer of epithelial cells of the preimplantation blastocyst embryo and all of its derivatives at later stages, comprising the giant cells and the majority of the fetal component of the placenta.

TS cells, trophoblast stem cells, stem cells derived from trophoblast of preimplantation or early postimplantation embryos that can be perpetuated in vitro as undifferentiated cells but have a developmental potential similar to trophectoderm, i.e., they can contribute to trophoblast derivatives in chimeras.

TUNEL, TDT-mediated UTP nick end labeling, method to visualize cells undergoing programmed cell death.

umbilicus, umbilical, cord-like structure, containing fetal veins and arteries, that connects the fetus to the placenta.

urogenital ridges, paired longitudinal ridges that form in the dorsal body wall; contain the mesonephroi and gonads.

visceral yolk sac, extraembryonic membrane composed of extraembryonic mesoderm and visceral endoderm.

vitelline, referring to the yolk sac. Vitelline circulation refers to the blood vessels in the yolk sac.

weaning, removal of pups from a lactating mother, usually done between 3 and 4 weeks of age.

wild type, an animal or allele that is considered normal and is the common type found in natural populations.

X-chromosome linked or X linked, a locus on the X chromosome.

X-gal, 5-bromo-4-chloro-3-indolyl-β-D-galactoside.

XO, referring to a mouse with only a single X chromosome and no Y chromosome; these animals are phenotypically female and fertile.

XX, referring to a mouse with two X chromosomes, normally a female.

XY, referring to a mouse with an X and a Y chromosome; normally a male.

ZP, zona pellucida, the thin, noncellular membrane, deposited around an oocyte during oogenesis, that the sperm penetrates during fertilization. It surrounds the preimplantation embryo until close to the time of implantation.

zygote, the fertilized oocyte.

Index

129-ES strain, 36, 38, 41, 53
2H3 antineurofilament monoclonal antibody, 136

A

AGM (aorta-gonad-mesonephros) region, 117
agouti, 38, 41
AI (artificial insemination), **215**
albino, 8, 38, 41, 43
Allantoic circulation, 77
Allantois, **215**
Alleles, **215**
Allele types
    coat color markers, 41, 42f, 43f
    complementation testing, 64, 65f
    conditional null, 26–28
    double-replacement strategy, 23, 25f
    floxed, 15, 18, 192
    flrted, 18, 192
    hypomorphic allele testing, 64
    naming, 49–50, 52f
    null, 18–19f, 19–20
    point mutations, 22–23, 24–25f
    reporter gene knock ins, 21–22
Allelic series, **215**
Alternative splicing, **215**
Anemia, 138, **215**
Aneuploid, 36, 42, **215**
Animal facilities, 50–51
Anlagen, 128, **215**
Antimorph, 5, **215**
Antimorph mechanism of action, 164, 165f
Antineurofilament monoclonal antibody, 136
Antiphosphohistone H3 antibody, 111
Aorta-gonad-mesonephros (AGM) region, 117, **215**
Apoptosis, 111

Arthritis, 153
Assisted reproduction
    female infertility, 63
    male infertility, 61–62
The Atlas of Mouse Development, 107
attPP' and *attBB'* system, 17
Autosomal, **215**

B

BAC (bacterial artificial chromosome), **215**
Backcross, **215**
Backcross breeding, 54, 55f
Background genotype, 51. *See also* Genetic background
Balance and motor coordination, 158–159
Balancer chromosomes, 60–61, 62–63f, **216**
Beta-galactosidase, 191, **216**
Bicistronic, 21, **216**
Blastocoel, 80, **216**
Blastocyst, 80, **216**
Blastocyst injection, 186–187
Blastocysts and implantation analysis
    failure to attach, 86, 87f
    failure to hatch, 86
    failure to outgrow, 88–89
Blastomeres, 80, **216**
Bleeding disorders, 138
Brain breathing center defects, 138
BrdU (bromodeoxyuridine), 109
Breeding
    backcross, 54, 55f
    schemes, 38
    techniques
        complementation testing, 64, 65f
        genetic interactions testing (*see* Genetic interactions testing)

Page numbers in bold indicate a glossary term.

Breeding (*Continued*)
    hypomorphic allele testing, 64
    increasing frequency of homozygous mutant mice, 69–70
    lethal phenotype workarounds, 194–195
Bromodeoxyuridine (BrdU), 109

C

C57BL/6 strains, 36, 38, 41, 52–54
Cardiac looping, 103
Cardiovascular defects
    perinatal lethality, 134, 135f
    prenatal lethality after 9.5 dpc, 119–121
Cauda epididymis, 148
Cell autonomous, 3, **216**
Cell death, 111, 127–128
Cell migration, 129
Central nervous system (CNS), 146, **216**
Chemical mutagens, 5–7
Chimeras, **216**
    breeding schemes for, 38
    dominant effects in, overview, 167–170
    dominant effects in heterozygous offspring
        causes of death, 171
        morphological or developmental abnormalities, 170–171, 172f
        reproductive problems, 171
    for lethal phenotype workarounds
        distinguishing cells, 179f, 180
        genotyping chimeras, 181
        information from chimeras, 176–177
        making a nonrandom mix of cells, 177–179
        making chimeras (*see* Chimeras, making)
        markers, 180, 182
    as markers, 180, 182
    morphological or developmental abnormalities, 170–171, 172f
    no chimeras produced analysis
        embryo death, 34
        maternal cannibalism, 33–35
        technique flaws, 32–33
    no offspring from male chimeras analysis, 37
    phenotype analysis examples and references, 205–206
    polarized, 187
    to study prenatal lethality, 130
    testing developmental potential in, 115
Chimeras, making, 177f
    basic methods and variables, 182
    checklist, 187
    developmental limitations, 182, 183
    extraembryonic tissue use, 182, 184
    lineage-restricted, blastocyst injection, 186–187
    lineage-restricted, morula aggregation, 185–186
    mixed-cell, 182, 184–185
    timing of analysis, 184

chinchilla, 41
Chi square test, 72–73, **216**
Chorioallantoic placenta, 75f, **216**
Chorion, 184, **216**
Chromosomes
    balancer chromosome use in maintaining a mutation, 60–61, 62–63f
    considerations in gene targeting, 3, 4f, 5
    genotyping for presence of Y, 40
    X inactivation, 5, 6f
    X-linked, 4f, 165, **226**
    Y-chromosome genotyping, 40, 165
CL (corpora lutea), 33, **216**
Cleavage, 80–84, 85f, **216**
Cleavage-stage aggregation, 182, 184–185
Cleft palate, 137–138
CNS (central nervous system), 146, **216**
Coat color markers, 41, 42f, 43f
Coat color phenotype, 36
Codominant, **216**
Co-isogenic strains, 38, **216**
Compaction, 74, **216**
Compact morula, 80
Complementation testing, 64, 65f, **217**
Compound heterozygote, 64, 67, **217**
Conditional alleles
    lethal phenotype workarounds, 188–190
    with linked reporter, 28
Conditional null alleles, 26–28
Congenic strains, 54, 55f, **217**
Construct, **217**
Corpora lutea (CL), 33, **216**
Cranial nerve defects, 134, 136
Cre, **217**
cre characterization, 190–195
Cre/*loxP* system, 17, 18
    generation of conditional null alleles, 26–28
    in point mutations, 22–23, 24f
Cre recombinase, 15
C-section, **217**
Cytospin, 118, **217**

D

Decidua, 122f, **217**
Decidualization, 89, **217**
Deleter mice, 59, **217**
Deletion complexes, 8–10
Developmental delay causing perinatal lethality, 134
Developmental Studies Hybridoma Bank, 136
Diagnostic restriction enzyme site, 16
Diapause, 90
Diaphragm defects, 138
DIC (differential interference contrast), 81, **217**
Diploid, 177, 179, 182, 183, **217**
Diploid morula, 185, 186

DNA
  building a targeting construct and, 14, 15f
  insertional mutagenesis and, 10–11
Documentation of embryos, 100
Documentation of mice, 146
Dominant, **217**
Dominant mutations
  in chimeras, 167–170
  dominant mutation defined, 163
  in ES cells, 166–167
  in heterozygous offspring of chimeras
    causes of death, 171
    morphological or developmental abnormalities, 170–171, 172f
    reproductive problems, 171
  imprinted genes and, 172–174
  isolating XO from XY ES cells, 170
  mechanism of action
    imprinted autosomal genes, 165
    overview, 163–165
    in X-linked gene, 165, 166f
    in Y-linked gene, 165
  phenotype analysis examples and references, 209
  transgene rescue, 168, 169f
Dominant negative mechanism of action, 164, 165f
Dominant sex reversal, 37, 46
Double heterozygote, 67, 168, **217**
Double-replacement strategy for alleles, 23, 25f
Downs and Davies staging system, 100–101f, 102
dpc (days postcoitus), **217**

E

Early lethal phenotype workarounds. *See* Lethal phenotype workarounds
Eating problems, 142
Electrofusion, 185, **217**
Electroretinogram (ERG), 156
Embryo proper, 183, **217**
Empty decidua, 77
Endochondral ossification, **218**
Endoreduplication, **218**
ENU (ethyl nitrosourea), 5–7, 8–9f, **218**
Environmental challenges to identify a mutant, 161
Epiblast, 59, 60, **218**
Epistasis, 65, **218**
ERG (electroretinogram), 156
Erythrocyte, **218**
ES cells, **218**
EST (expressed sequence tag), 20, **218**
Estrogen receptor, Cre, 193
Estrous cycle, 39
Ethyl nitrosourea (ENU), 5–7, 8–9f, **218**
Euploid, 36, **218**
Exon checklist, 14
Expressed sequence tag (EST), 20, **218**
Expressivity, 51, 66, **218**

Extraembryonic, 3, 4f, 184, **218**
Extraembryonic ectoderm, **218**

F

Facilities, animal, 50–51
FACS (fluorescence-activated cell sorting), 128, **218**
Female mice
  corpora lutea counting, 33
  estrous females selection, 39
  growth defects, 143
  infertility, dealing with, 63
  infertility diagnosis, 150–151
  maternal cannibalism issues, 33–34, 35
  sex identification, 46, 47f
  sexing embryos, 92, 93f
  X-chromosome-linked gene mutation, 4f, 165, 166f
Floxed, **218**
Floxed allele, 15, 18, 192
Floxed selection cassette, 59–60
Flp/*FRT* system, 17
Flrted, **218**
Flrted allele, 18, 192
Fluorescence-activated cell sorting (FACS), 128, **218**
Food, soft, 147
Forced heterozygosity, **218**

G

G418, 15
Gait analysis, 159, 160f
Gametes, 6, **218**
Gametogenesis, 16, **218**
Gastrointestinal defects, 148
Gene expression
  cell-type-specific analysis, 108
  overexpression and mutant phenotypes, 68–69
  pattern considerations in targeting, 2–3
  prenatal lethality analysis and, 94, 108, 127
  quantitative assessments, tissue-specific, 127
Gene targeting, **218**
Gene targeting in ES cells
  clonal lines isolation (*see* Phenotype recognition and absence troubleshooting)
  experiment steps, 1–2
  gene checklist, 3, 4f, 5
  gene expression pattern considerations, 2–3
  phenotype analysis examples and references, 205
  project planning steps, 2
Gene-targeting strategies
  building a targeting construct
    checklist, 14
    essential elements, 15f
    homologous DNA use, 14, 15f
    negative selection, 16
    positive selection, 15–16
    screening for homologous recombination events, 16–17

Gene-targeting strategies (*Continued*)
    selection cassette removal, 17
    site-specific recombinases, 17–18
  chimeras production and breeding, 32, 37, 45
  conditional null alleles and, 26–28
  null alleles and, 18–19f, 19–20
  phenotype analysis examples and references, 205–206
  point mutations, 22–23, 24–25f
  replacement vector use, 20
  reporter gene knock ins, 21–22
Genetic background, **219**
Genetic challenges to identify a mutant, 160–161
Genetic interactions testing
  biochemical pathway components, 66–68
  mutant phenotypes caused by overexpression of downstream genes, 68–69
  overlapping expression patterns, 68
  overview, 65–66
  redundancy or compensation, 66, 67f
  similar mutant phenotypes caused by unrelated genes, 68, 69f
Genetic modifier, **219**
Gene traps, 10, **219**
Genotype, **219**
Germ layers, **219**
Germ-line transmisson, **219**
Gestation, **219**
GFP (green fluorescent protein), 28, **219**
Giant cells, 72, **219**
Growth defects, 143

H

Haploid, 44, 45, **219**
Haploid effect on sperm, 45
Haploinsufficiency, **219**
  lethal mutation from, 70
  likely results of, 167
  mechanism of action, 163, 165f
Hatching, 75f, 81, **219**
Hearing problems, 142
Hearing test, 156–157
Heart examination, 99, 102, 103. *See also* Cardiovascular defects
Hematopoiesis, **219**
Hematopoietic defects
  anemia, 138
  bleeding disorders, 138
  prenatal lethality after 9.5 dpc, 116–119
Hemizygous, 32, **219**
Herpes simplex virus (HSV), 16, 42
Herpes simplex virus thymidine kinase (HSV *tk*), 10, 16, 17
Heterozygote, 22, **219**
Heterozygous x wild-type cross, 56
Histocompatible, **219**

Histone-H2-eGFP, 83, 85f
Hit-and-run strategy for alleles, 23, 25f
Homolog/homologous, **219**
Homologous DNA, **219**
Homologous recombination, **220**
Homozygote, **220**
Hybrid mice, **220**
Hybrid vigor, **220**
Hypermorph, **220**
Hypermorph mechanism of action, 163, 165f
Hypomorph, **220**
Hypomorphic allele testing, 64

I

IACUC (Institutional Animal Care and Use Committee), 55–56
ICM (inner cell mass), 80, **220**
ICSI (intracytoplasmic sperm injection), 62, 63, **220**
Immunosurgery, **220**
Implantation, 34, **220**
Implantation delay, **220**
Imprinting, **220**
  autosomal genes, 165
  dominant effects and, 172–174
  inheritance of mutant genes, 6f
  X inactivation and, 5, 6f
Inbred strains, 14, **220**
Inbreeding, **221**
Inbreeding depression, 56–57, **221**
Infertility. *See also* Reproduction problems
  female, 63, 150–151
  male, 16, 61–62, 147–150
Inner cell mass (ICM), 80, **220**
Insertional mutagenesis, 10–11
Insertional mutation, 22, **221**
Institutional Animal Care and Use Committee (IACUC), 55–56
Intercross, 43, **221**
Intersex chimeras, 37
Intracytoplasmic sperm injection (ICSI), 62, 63, **220**
IRES (internal ribosome entry site), 21, **221**
Isogenic, **221**
IVF (in vitro fertilization), 61–62, **221**

K

Karyotype, 42, **221**
Kaufmann, Matt, 107
Knock ins
  characterizing, 190–195
  reporter gene, 21–22
Knockout mice, 28, **221**
Kyphosis, 153

L

Labyrinth, 122, 123, **221**
lacZ, 21f, 28

Lethal phenotype workarounds
  breeding techniques, 194–195
  chimeras use
    distinguishing cells, 179f, 180
    genotyping chimeras, 181
    information from chimeras, 176–177
    making a nonrandom mix of cells, 177–179
    making chimeras (*see* Chimeras, making)
    markers, 180, 182
  conditional alleles, 188–189
    testing, 189
    using, 190
  phenotype analysis examples and references, 209
  recombinase-expressing transgenic mouse lines, 190–195
  regulatory mutations creation, 195–196
  transgene rescue, 188
Locomotion abnormalities, 145–146
Locus, **221**
Loss of function, **221**
loxP, 15, 17–18, **221**
  Cre/*loxP* system, 17, 18
    generation of conditional null alleles, 26–28
    in point mutations, 22–23, 24f

M

Male mice
  chimeras' inability to produce offspring, 37
  ENU mutagenesis effects, 7
  growth defects, 143
  infertility, dealing with, 61–62
  infertility diagnosis, 147–150
  isolating XO from XY ES cells, 170
  sex identification, 46, 47f
  sexing embryos, 92, 93f
  sterility due to HSV *tk*, 16
  Y-chromosome-linked gene mutation, 165
Marker recycling, 16, **221**
Markers
  coat color, 41, 42f, 43f
  to document cell migration, 129
  inactive selectable, 31–32
  for molecular characterization of mutants, 107–109
  nuclear-localized transgenic, 83, 85f
  pigmentation, 182
  for positive selection, 15
  using chimeras as, 180, 182
MBP (myelin basic protein), 146
MEFs (mouse embryonic fibroblasts), 128, **221**
Megacolon, 148
Meiosis, 38, **221**
Mendel's laws, 69–70, 72–73, 206
Mesonephros, **221**
MI (mitotic index), **222**
Moles, 77
Moribund, **222**

Morula, **222**
Morula aggregation, 182, 184–186
Mosaic, 3, 17, **222**
Mouse embryonic fibroblasts (MEFs), 128, **221**
Mouse Genome Informatics (MGI) Phenotype Ontology Browser, 144–146
Mutant allele, **222**
Mutants maintenance
  animal facilities, 50–51
  backcross breeding, 54, 55f
  balancer chromosome use, 60–61, 62–63f
  breeding techniques
    complementation testing, 64, 65f
    genetic interactions testing (*see* Genetic interactions testing)
    hypomorphic allele testing, 64
    increasing frequency of homozygous mutant mice, 69–70
  female infertility, dealing with, 63
  genetic background effects, 51–54
  inbred strain choice/allele transfer to, 52–54
  inbreeding depression avoidance, 56–57
  male infertility, dealing with, 61–62
  naming a mutant, 49–50, 52f
  phenotype analysis examples and references, 206
  production colony maintenance, 55–57
  production of mutant mice or embryos, 57
  selection cassette deletion in vivo, 58–60
Mutation, **222**
Mutations induction
  chemical mutagens, 5–7, 8–9f
  gene-targeting basics (*see* Gene targeting in ES cells)
  insertional mutagenesis, 10–11
  phenotype analysis examples and references, 205
  transgenics, 11
  X-ray mutagenesis, 7–10
Myelin basic protein (MBP), 146
Myelin disfunction, 146
Myelin proteolipid protein (PLP), 146

N

Naming a mutant, 49–50, 52f
Necropsy guide for tissue collection, 153
Necrotic cell death, 111
neo (neomycin phosphotransferase), 15–16, **222**
neo expression cassette, 24–25f, 26–27f
Neomycin, **222**
Neomorph mechanism of action, 164, 165f
Neural tube closure, 103, 104f, 105
Neurological problems, 144–146
Neurological tests
  balance and motor coordination, 158–159
  gait analysis, 159, 160f

Neurological tests (*Continued*)
  strength, 159
  swim test, 159
  tail suspension test, 158
Nociception, 157
Nuclear-localized transgenic markers, 83, 85f
Null, **222**
Null alleles
  conditional, 26–28
  making, 19–20
  mechanism of action, 163, 164–165

O

OCT, 108, **222**
Olfaction problems, 142
129-ES strain, 36, 38, 41, 53
Oogenesis, 42, 46, 59, **222**
Oogonia, **222**
ORF (open reading frame), **222**
Organogenesis, 75f, 124, **222**
Ortholog/orthologous, **222**
Orthotopic, **222**
Osteopetrosis, 153
Osteoporosis, 153
Outbred or noninbred, 36, **222**
Outcross, 54, **223**
Ovulation
  corpora lutea counting and, 33
  estrous cycle, 39

P

Paraformaldehyde (PFA), **223**
Parietal yolk sac, 121, **223**
Parturition, 34, 35, **223**
PCNA (proliferating cell nuclear antigen), **223**
PCR (polymerase chain reaction), 7, 16
PCR genotyping, 40
Pde6b (retinal degeneration), 157
PECAM (platelet endothelial cell adhesion molecule), 121
Penetrance, **223**
Percardial effusion, 119f, **223**
Perinatal, **223**
Perinatal lethality analysis
  anemia, 138
  bleeding disorder, 138
  brain breathing center defects, 138
  cardiovascular defects, 134, 135f
  catastrophic abnormalities, 134
  cleft palate, 137–138
  cranial nerve defects, 134, 136
  determining time of death, 132–133
  developmental delay, 134
  diaphragm defects, 138
  possible causes, 132
  skeletal defects, 136, 137f

Peripheral nervous system (PNS), 146
PFA (paraformaldehyde), **223**
pgk gene, **223**
Phenotype, **223**
Phenotype analysis examples and references
  chimeras, 205–206
  dominant mutations, 209
  gene-targeting strategies, 205–206
  lethal phenotype workarounds, 209
  Mendel's laws, 206
  mutants maintenance, 206
  mutations induction, 205
  postnatal mutant analysis, 208–209
  prenatal lethality, 206–208
Phenotype recognition and absence troubleshooting
  absence of heterozygous offspring, 45–46
  adequacy of cell production, 31
  breeding schemes for chimeras, 38
  coat color markers, 41, 42f, 43f
  common mistakes in gene-targeting experiments, 30
  estrous females selection, 39
  genotyping for presence of Y chromosome, 40
  haploid effect on sperm, 45
  insufficient recovery and growing of targeted ES cells
    heterozygous mutant effect, 32
    inactive selectable marker, 31–32
    possible technical difficulties, 30
    targeting construct trouble, 30–31
  no chimeras produced
    embryo death, 34
    maternal cannibalism, 33–35
    possible technique flaws, 32–33
  no germ-line transmission, 40, 42
  nonchimeric pups produced, 34–36
  no offspring from male chimeras, 37
  number of oocytes ovulated determination, 33
  sex identification, 46, 47f
  sufficiency of numbers of cells produced, 31
  wild-type allele only recovered, 43–45
Photographic documentation of embryos, 100, 125
Pigmentation as a marker, 182
pink-eyed dilution, 41, 43
Placental insufficiency, 121–123
Platelet endothelial cell adhesion molecule (PECAM), 121
PLP (myelin proteolipid protein), 146
PNS (peripheral nervous system), 146, **223**
Point mutations, 22–23, 24–25f
Polarized chimera, 187
Polymerase chain reaction (PCR), 7, 16
Polymorphic, **223**
Postnatal, **223**
Postnatal mutant analysis
  marking newborn pups, 139

no visible mutant phenotypes, 154
  balance and motor coordination, 158–159
  environmental challenges, 160–161
  gait analysis, 159, 160f
  genetic challenges, 160–161
  hearing test, 156–157
  possible causes, 155
  smell and taste test, 157
  strength, 159
  swim test, 159
  tail suspension test, 158
  touch test, 157–158
  vision test, 155–156
perinatal lethality
  anemia, 138
  bleeding disorder, 138
  brain breathing center defects, 138
  cardiovascular defects, 134, 135f
  catastrophic abnormalities, 134
  cleft palate, 137–138
  cranial nerve defects, 134, 136
  developmental delay, 134
  diaphragm defects, 138
  possible causes, 132
  skeletal defects, 136, 137f
  time of death determination, 132–133
phenotype analysis examples and references, 208–209
phenotype analysis technique, 138–140, 141f
video documentation of mice, 146
visible mutant phenotypes, age-dependent
  approach to studying, 153–154
  necropsy guide for tissue collection, 153
  skeletal defects, 153–154
  tumor formation, 151–153
visible mutant phenotypes before/after weaning/sexual maturity (see Weaning)
Pregnancies. See also Prenatal lethality analysis; Reproduction problems
  estrous cycle, 39
  maternal cannibalism issues, 33–34, 35
  ovulation, 33, 39
Preimplantation development, 81f
Preimplantation phenotype analysis
  abnormal blastocyst morphology, 85–86
  arrest or delay during cleavage, 83, 84, 85f
  cell counting techniques, 84
  of cleavage stages and early blastocysts, 80–83
  no compaction, 85
  no mutant embryos, 83
Prenatal lethality after 9.5 dpc
  cardiovascular defects, 119–121
  causes of death during period, 116
  developmental defects in morphogenesis and organogenesis
    cell proliferation and death, 127–128
    chimeras use, 130

documentation of defect development, 124
    fundamental questions to ask, 124
    histological analysis, 124–125
    quantitative assessments of tissue-specific gene expression, 127
    skeletal analysis, 125, 126f
    specimen orientation considerations, 125
    in vitro analysis of cells and organs, 128–129
  hematopoietic defects, 116–119
  molecular characterization of mutants, 127
  placental insufficiency, 121–123
Prenatal lethality analysis
  between 4.5 and 9.5 dpc
    at 5.5 dpc, 96
    between 6.5 and 9.5 dpc (see Prenatal lethality between 6.5 and 9.5 dpc)
    developmental landmarks after 4.5 dpc, 78
    between gastrulation and allantoic fusion, 99
    lethality in early gestation indications, 94
    orientation of embryos, 97, 98f
    periimplantation death, 4.5–5.5 dpc, 94–96
    postimplantation stages, 95f
  blastocysts and implantation analysis
    failure to attach, 86, 87f
    failure to hatch, 86
    failure to outgrow, 88–89
  Chi square test, 72–73
  cleavage and blastocyst formation process, 82–83
  control levels of embryonic and fetal loss establishment, 79
  embryonic and fetal development schematic
    after 4.5 dpc, 78
    up to 12.5 dpc, 75f
  if homozygous mutants not present at birth, 72, 74
  if homozygous mutants present at birth, 71–72
  implantation delay analysis, 90–91
  isolated ICM analysis, 89–90
  midgestation to late-gestation phenotypes after 9.5 dpc (see Prenatal lethality after 9.5 dpc)
  morphological analysis of preimplantation embryos
    categories for sorting, 80, 81f
    gene expression, 94
    genotype determination, 91
    histology, 92–93
    sexing the embryos, 92, 93f
  number of mutants to examine determination, 124
  phenotype analysis examples and references, 206–208
  photographic documentation of embryos, 100
  preimplantation development, 81f
  preimplantation phenotype analysis
    abnormal blastocyst morphology, 85–86
    arrest or delay during cleavage, 83, 84, 85f
    cell counting techniques, 84
    of cleavage stages and early blastocysts, 80–83

234 | Index

Prenatal lethality analysis (*Continued*)
   no compaction, 85
   no mutant embryos, 83
   time of death assessment
      allantoic circulation establishment, 77
      analysis after 12.5 dpc, 74–77
      analysis between 4.5 and 9.5 dpc, 77
      lethality indications before 4.5 dpc, 77–78, 80
      in vitro culture of embryos, 82
      zona pellucida recovery/removal, 80, 86, 87f, 88
Prenatal lethality between 6.5 and 9.5 dpc
   cell proliferation and death
      measuring cell death, 111–112
      measuring cell proliferation, 109–111
      usefulness of analysis, 109
   cell-type-specific gene expression analysis, 108
   common causes, 99
   developmental potential, 112
      making cell lines, 114–115
      teratomas from transplanted embryonic fragments, 113–114
      testing in chimeras, 115
      in vitro culture of cells, tissues, and organs, 113
      in vitro culture of postimplantation embryos, 112–113
   developmental stage determination, 99, 100–101f, 102, 103
   embryo histological assessment, 105, 106f
   fixing and embedding later-stage embryos, 105, 107
   between gastrulation and allantonic fusion, 97
   genotyping embryos, 107
   heart examination, 99, 102, 103
   molecular characterization of mutants, 107–109
   morphological assessment importance, 99, 100
   neural tube closure, 103, 104f, 105
Progesterone receptor, Cre, 193
Proliferating cell nuclear antigen (PCNA), **223**
Pronucleus, 81f, **223**
Provirus, **223**
Pseudopregnant, 32, **223**

R

RBCs (red blood cells), **223**
Recessive, **223**
Recombinase-expressing transgenic mouse lines, 190–195
Red blood cell development, 118f
Red blood cells (RBCs), **223**
Redundancy, **223**
Regulatory mutations, 195–196
Reproduction problems
   female infertility, 63
   male infertility, 61–62
   ovulation

   corpora lutea counting and, 33
   estrous cycle, 39
   placental insufficiency, 121–123
   problems in heterozygous offspring of chimeras, 171
Rescuing transgene, **223**
Resorption or resorption site, 34, 77, **223**
Resources, books, 97–202
Resources, web-based, 202–204
Retinal degeneration (*Pde6b*), 157
Retroviruses, 10, **224**
Rosa26, 184, **224**
RU-486, 193

S

Selection cassette removal
   in vitro, 17
   in vivo, 58–60
Semidominant, **224**
Sex identification, 46, 47f, 92, 93f
Site-specific recombinases, 17–18
Skeletal defects
   age-dependent, 153–154
   causing perinatal lethality, 136, 137f
   causing prenatal lethality, 125, 126f
Sleeping Beauty (SB) transposase, 10–11
Smell and taste test, 157
Somite number, 99, 102, 103f
Southern analysis, 16
Specific pathogen-free (SPF) barrier, 50–51, **224**
Speed congenic, 54, **224**
Spermatogenesis, **224**
Spermatogenesis and infertility in males, 148–150
Spermatogonia, **224**
Spermiogenesis, **224**
SPF (specific pathogen free), 50–51, **224**
Spongiotrophoblast, 122, **224**
Sterility in males, 16, 61–62, 147–150
Stock, **224**
Strength test, 159
Substrain, **224**
Superovulation, 61, 63, **224**
Swim test, 159
Syncytiotrophoblast, **224**
Syntenic, **224**

T

Tail-flick test, 157
Tail suspension test, 158
Tamoxifen, 193
Targeted mutagenesis, **224**
Targeting construct building
   checklist, 14
   homologous DNA use, 14, 15f
   negative selection, 16

positive selection, 15–16
screening for homologous recombination events, 16–17
selection cassette removal, 17
site-specific recombinases, 17–18
Targeting strategies. *See* Gene-targeting strategies
Teeth, 147
TEM (transmission electron microscopy), **224**
Teratocarcinoma, **224**
Teratoma, 113–114, **224**
Terminal dUTP nick end labeling (TUNEL), 111, 128, **225**
Term or full term, **225**
Tetraploid embryos, 177, 185, 186, **225**
Thymidine kinase (*tk*), 16, 42
Tissue-specific loss-of-function alleles, 195–196
Totipotency, 176, **225**
Touch test, 157–158
Transgene, **225**
Transgene rescue
of a dominant lethal effect, 168, 169f
lethal phenotype workarounds, 188
Transgenic mice, **225**
Transgenics, 11
Transmission electron microscopy (TEM), **224**
Transposase, **225**
Transposons, 10–11, **225**
Trophectoderm, **225**
Trophectoderm cells, 80, 185
Trophoblast, **225**
Trophoblast giant cells, **225**
Trophoblast stem (TS) cells, 115, **225**
Tumor formation, 151–153
TUNEL (terminal dUTP nick end labeling), 111, 128, **225**
2H3 antineurofilament monoclonal antibody, 136
Tyr^c, 41, 43

U

Umbilicus, umbilical, **225**

V

Vasculogenesis, 120
Vestibular system, 146
Video documentation of mice, 146
Viruses, 10
Visceral yolk sac, 121, **225**
Vision test, 155–156
Visual tracking drum, 156
Vitelline, 103, **225**
Vocalizations of mice, 142
Von Frey filament test, 157

W

Weaning, **225**
importance of, 140–141
visible mutant phenotypes after weaning/sexual maturity
compromised growth and vigor, 147–148
infertility in females, 150–151
infertility in males, 147–150
visible mutant phenotypes before weaning
altered growth pattern, 143
locomotion abnormalities, 145–146
morphological abnormalities, 143–144
neurological problems, 144–146
preweaning lethality causes, 142
white-bellied agouti, 41
Wild type, **226**

X

X-chromosome-linked gene mutation imprinting, 4f, 165, 166f
X-chromosome linked or X linked, **226**
X-gal, 22, **226**
X inactivation, 5, 6f
XO, **226**
X-ray mutagenesis, 7–10
XX, **226**
XY, **226**

Y

Y-chromosome genotyping, 40, 165
Yolk sac, visceral, **225**
Yolk sac and prenatal lethality, 121

Z

Zona pellucida (ZP), 80, 87f, 88, **226**
Zygote, 10, **226**